Q
175.32
.I54
S73
2006

DISCARD

Exceeding Our Grasp

Exceeding Our Grasp

Science, History, and the Problem of Unconceived Alternatives

P. KYLE STANFORD

2006

OXFORD
UNIVERSITY PRESS

Oxford University Press, Inc., publishes works that further
Oxford University's objective of excellence
in research, scholarship, and education.

Oxford New York
Auckland Cape Town Dar es Salaam Hong Kong Karachi
Kuala Lumpur Madrid Melbourne Mexico City Nairobi
New Delhi Shanghai Taipei Toronto

With offices in
Argentina Austria Brazil Chile Czech Republic France Greece
Guatemala Hungary Italy Japan Poland Portugal Singapore
South Korea Switzerland Thailand Turkey Ukraine Vietnam

Copyright © 2006 by Oxford University Press, Inc.

Published by Oxford University Press, Inc.
198 Madison Avenue, New York, New York 10016

www.oup.com

Oxford is a registered trademark of Oxford University Press.

All rights reserved. No part of this publication may be reproduced,
stored in a retrieval system, or transmitted, in any form or by any means,
electronic, mechanical, photocopying, recording, or otherwise,
without prior permission of Oxford University Press.

Library of Congress Cataloging-in-Publication Data
Stanford, P. Kyle.
Exceeding our grasp: science, history, and the problem of unconceived
alternatives / P. Kyle Stanford.
p. cm.
Includes bibliographical references and index.
ISBN-13 978-01-19-517408-3
ISBN 0-19-517408-9
1. Inference. 2. Reasoning. 3. Science—Philosophy. I. Title
Q175.32.I54S73 2006
501—dc22 2005051299

9 8 7 6 5 4 3 2

Printed in the United States of America
on acid-free paper

For Alyssa and Casey,
without whom not

Preface

Though I didn't know it at the time, I began writing this book during a conversation with Jessica Pfeifer that she probably does not remember, soon after we had both entered graduate school in the Philosophy Department and Science Studies Program at the University of California, San Diego. After I grandly and self-importantly explained the problem of underdetermination, Jessica replied in her characteristically self-effacing and commonsensical way that she could understand how the available evidence might support two alternative theories equally well in principle, but she asked why she should believe that there really are always (or ever) such equally well-supported alternatives to our own scientific theories out there to be found. She hastened to add, as she invariably does, that her confusion was almost certainly caused by her inability to understand what I was saying rather than the fault of any infelicity in my expression or any underlying problem with the view I was espousing. But this was simply not so. I did not have a good answer to Jessica's question at the time, and in trying to figure out the right answer to it I have been led to reconceive not only what I think the most serious problem about underdetermination really is but also virtually everything else I think is at the heart of the issue of scientific realism. I have no reason to think Jessica will be any more impressed with the answer I have now than the slack-jawed blank stare I gave her in 1991, but I am very glad that she asked me the question.

While there are a very large number of others to whom I owe intellectual debts both in connection with this book and more generally, there are a small handful to whom I must express my deepest gratitude for what I regard as their incredible generosity and apparently endless willingness to devote time and energy to talking to me both about the arguments I develop below and about related philosophical subjects. My friend and former graduate advisor Philip Kitcher as well as three of my colleagues here at the University of California, Irvine (UCI)—Jeff Barrett, Pen Maddy, and David Malament—have quite literally been with the project of this book every step of the way. Each has demonstrated an astonishing willingness to thoughtfully read, and reread, and talk to me about every chapter and virtually every issue discussed in the pages that follow. They have helped in ways large and small—from making sensible suggestions about strategy and proposing useful examples to refusing to accept my unconvincing answers to their skeptical questions, and I quite literally cannot imagine what this book would now say were it not for their uniformly constructive and generous input. Each of these friends and colleagues has been willing to divert significant amounts of time and effort from their own projects, interests, and professional careers in order to advance my own. I do not hold out any hope of ever managing to repay their kindness.

Other colleagues have also been willing to read and discuss substantial parts of the manuscript, and I am extremely grateful for their useful feedback and thoughtful discussions. Peter Godfrey-Smith and Stathis Psillos each reviewed the entire project in an embryonic form and provided invaluable advice. I am deeply grateful as well to others for constructive comments and productive discussions concerning particular arguments and specific parts of the manuscript in earlier forms, especially Arthur Fine, Jarrett Leplin, John Norton, Bob Batterman, Hasok Chang, Anjan Chakravartty, Greg Radick, Peter Lipton, Ludwig Feurbach, Alan Nelson, Larry Laudan, Alexander Rosenberg, Brian Skyrms, Craig Callender, Bill Demopoulos, Elliott Sober, Sandy Mitchell, Bas Van Fraassen, Aldo Antonelli, Sherri Roush (did I spell it right, Sherri?), Kim Sterelny, Noretta Koertge, Baron Reed, Larry Sklar, Andre Kukla, P. D. Magnus, James Ladyman, Jonathan Hodge, John Worrall, Tim Lyons, Kent Staley, Jim Lennox, Paul Griffiths, Otavio Bueno, Jessica Poulin, Francisco Ayala, Robert Olby, and John Earman. I am grateful as well to the many people on this list who were willing to discuss their work with me in private correspondence, sometimes at considerable length. And a special debt of thanks is owed to Anjan Charkravartty, who was also kind enough not only to stand in for me to present some of this work at a meeting of the British Society for the Philosophy of Science when my son Casey was born rather suddenly and unexpectedly, but also to try to suggest the responses to objections that he supposed I might give—a service that was surely above and beyond the call of duty.

I am deeply grateful as well to a number of historians of science with whom I have consulted in one way or another about the historical evidence I present in this work, including Jane Maienschein, Peter Bowler,

Mary Terrall, Rasmus Winther, Martin Rudwick, and especially Keith Benson. Here it is especially important to be clear, however, that the willingness of these scholars to provide helpful suggestions and comments should not be confused in any way with sympathy for even my most basic contentions. My thanks also go to Nick Jolley, Alan Nelson, and especially David Lemoine for constructive advice on my translations of scientific writings from their original sources. I am also grateful to the Cambridge University Library and to the Staatsbibliothek zu Berlin—Preußischer Kulturbesitz for providing me with copies of unpublished letters from Charles Darwin to various correspondents (see chapter 3).

Earlier versions of the arguments presented here also benefited considerably from suggestions by anonymous reviewers for *Philosophy of Science, The Journal of Philosophy, Biology and Philosophy*, Oxford University Press, and for my grant proposal to the National Science Foundation (thanks are also due to a series of extremely helpful and encouraging program officers at the Science and Technology Studies Program of the NSF). I have also benefited from discussing this material with audiences at various conferences, including meetings of both the Philosophy of Science Association and the British Society for the Philosophy of Science (on multiple occasions each), the Canadian Society for the History and Philosophy of Science, the British Society for the History of Science, and the Southern California Philosophy Conference, as well as with audiences at Stanford University, the Claremont Colleges Colloquium (Pomona College), the University of Missouri, and UC Santa Cruz.

I have had the good fortune to be in constant contact with bright, enthusiastic, articulate, and insightful graduate students both here at UCI and elsewhere throughout the time that I developed the arguments presented in this book: they have made me think harder, longer, and better about the central issues. I am especially indebted in this regard to Kevin Zollman, Carol Skrenes, Brian Woodcock, Patrick Forber, Patricia Marino, Arash Pessian, Christina McLeish, Teri Merrick, Will Rowley, Waldemar Rohloff, Jason Alexander, Jason Ford, and Doug Hill, as well as a number of others who attended a variety of graduate seminars I have offered in recent years concerning the subjects addressed in the pages that follow.

I would also like thank Peter Ohlin, my editor at Oxford University Press, for his unfailing encouragement, patience, and good sense, and OUP itself for its willingness to be flexible with me about contracts and deadlines. I also owe my thanks and formal acknowledgment to several other publishers for granting me permission to revise and republish parts of my own previously published work as follows:

Chapter 1 incorporates material from two previously published papers. First, "An Antirealist Explanation of the Success of Science," *Philosophy of Science* 67: 266–284. © 2000 by the Philosophy of Science

Association. All rights reserved. And secondly, "Refusing the Devil's Bargain: What Kind of Underdetermination Should We Take Seriously?," *Philosophy of Science 68 (Proceedings):* S1-S12. © 2001 by the Philosophy of Science Association. All rights reserved.

Chapter 6 is based on "Pyrrhic Victories for Scientific Realism," *The Journal of Philosophy* 100: 553–572. © 2003 by The Journal of Philosophy, Inc. All rights reserved.

Chapter 7 is based on "No Refuge for Scientific Realism: Selective Confirmation and the History of Science," *Philosophy of Science* 70: 913–925. © 2003 by the Philosophy of Science Association. All rights reserved.

Chapter 8 incorporates small parts of my entry on "Instrumentalism" in Sahotra Sarkar and Jessica Pfeifer (eds.) *The Philosophy of Science: An Encyclopedia* Routledge: New York. © 2005 by Taylor and Francis, Inc. All rights reserved.

I am also grateful to the National Science Foundation, the School of Social Sciences, and the School of Humanities here at UCI, and to my own campus and the University of California as a whole for providing significant financial support for my research on this project. I would like to thank both the School of Social Sciences and the Program in History and Philosophy of Science at UCI (thanks Brian!) for supporting travel related to my research for this project and presentation of parts of it to various professional conferences. I am grateful to UCI for supporting this work with a Career Development Award and sabbatical release, and to the Department of Logic and Philosophy of Science (as well as the Philosophy Department) for providing a stimulating and supportive environment for research with wonderful colleagues, graduate students, and friends. Some of this material is based upon work supported by the National Science Foundation under Grant No. SES—0094001. Any opinions, findings, and conclusions or recommendations expressed in this material are those of the author and do not necessarily reflect the views of the National Science Foundation.

I have also benefited in ways that I cannot hope to count or disentangle from the specific project of this book from many conversations over the years concerning scientific realism and the alternatives with a large number of other friends and colleagues including: Peter Kosso, Jessica Pfeifer, Bruce Glymour, Andrew Wayne, Martin Rudwick, Jonathan Gunderson, Gillian Barker, Josh Jorgensen, John Dupre, Brian Keeley, Alan Richardson, Aarre Laakso, Alyssa McIntyre Stanford, Steve Yalowitz (especially regarding Bunsen burners), Paul Churchland, Pat Churchland, Pat Kitcher, Richard Arneson, Gila Sher, Jack Bradbury, Sandy Vehrencamp, Paul Harris, Josh Kohn, Steven Epstein, and many other residents and visitors in both the Philosophy Department and the Science Studies Program at UC San Diego, as well as the Departments of Philosophy and Logic and Philosophy of Science at UCI. This does not by any means, however, exhaust the list of people who have been helpful to me in writing this book. It is perhaps equally important to acknowledge the innumerable sparks of insight and thought-provoking remarks provided to me by others in casual conversations enjoyed purely out of shared intellectual interest in hallways, hotel elevators, taxicabs, airports, seminar rooms, on playgrounds, at poker games, in pool halls, on basketball courts, and (of course) in bars. The sheer volume of generous help I have received ensures that my memory will be inadequate to acknowledge it fully here, and I can only apologize in advance to those whom I have inexcusably forgotten and hope that my inattention will not be interpreted as ingratitude.

At different times, Darwin wrote that the theory of Pangenesis was both a "mad dream" and was his "beloved child." I now think I know what he meant. In writing this book, I have been through periods of being quite sure that the arguments and evidence I have marshaled could not hope to convince anyone of anything, and periods of being equally sure that the book need only see the light of day to start winning hearts and minds. I am content to hope that the truth lies somewhere in between. Moreover, I doubt that this will be the last word I ever write about the subjects discussed in the pages that follow. Today I am sure that the view advanced in this book represents my own best-considered judgment on the issues it addresses. But today is a good day.

Darwin's work also interpenetrated every facet of his life at home with his wife and children. As a famous story goes, in the midst of the twelve years that Darwin spent working on barnacles, he took one of his children to a friend's house for tea. The child looked around the house in a puzzled way for a while and then proceeded to inquire of one of the children of his host, "and where does your father do *his* barnacles?" I am afraid that every academic family knows what it must have been like for Darwin's. For their constant indulgence, their infinite patience, and their unfailing encouragement, I owe another debt of gratitude that cannot be repaid to my own actual beloved child, Casey, and to my wife Alyssa, who has made my own life a dream, though one that is mad only in the sense that I am struck every day by my improbable and incredible good fortune in finding her, in making a life together with her, and in her continued willingness to put up with me. In many more ways than I suspect they realize, this book is for them.

Contents

Exceeding Our Grasp

1

Realism, Pessimism, and Underdetermination

Theories come and theories go. The frog remains.

—Jean Rostand, *Notebooks of a Biologist*

1.1 Scientific Realism: What's at Stake?

Suppose you crack open a textbook on particle physics, evolutionary biology, molecular genetics, or just about any scientific subject—what will you find there? No matter the field, at the heart of the text you will typically find a theory purporting to describe some part or aspect of the world that has been the subject of systematic investigation by human beings. Typically, the domains described by our scientific theories have been the subject of devoted and systematic theoretical investigation precisely because it is difficult to acquire information about them otherwise. This might be because the entities inhabiting them are too small or too large or too amorphous for us to readily perceive; because the causal interactions between those entities are too fast or too slow or too rare or take place on too grand a scale for us to engage with in ordinary ways; these entities and interactions occur in times and places either far removed from our own or otherwise inconveniently located (e.g., at the dawn of life on Earth, in remote regions of the universe, at the center of the Sun), and so on. But along with the textbook presentation of such a scientific theory you will also find at least some of the evidence that has led to its acceptance, including a description of what that theory has enabled us to explain, predict, and/or accomplish. In other words, in a scientific textbook you will typically find a story about how things stand in some otherwise largely inaccessible domain of nature, along with at least some accounting of why we think the story is true and of the practical and explanatory achievements made available to us by thinking of the world as really being the way that the story says it is. A fact-finding trip to the university bookstore at the University of California Irvine where I work produced the following representative examples of theoretical claims from some of the textbooks that were used in science courses in a recent spring term:

> In ionic bonding the participating atoms are so different that one or more electrons are transferred to form oppositely charged ions, which then attract each other. In covalent bonding two identical atoms share electrons equally. The bonding results from the mutual attraction of the two nuclei for the shared electrons. Between these extremes are intermediate cases in which the atoms are not so different that electrons are completely transferred but are different enough so that unequal sharing results, forming what is called a polar covalent bond. An example of this type of bond occurs in the hydrogen fluoride (HF) molecule. When a sample of hydrogen fluoride gas is placed in an electric field, the molecules tend to orient themselves . . . with the fluoride end closest to the positive pole and the hydrogen end closest to the negative pole. (Zumdahl and Zumdahl 2003 350–351)

> Through intensive research efforts over the past twenty-five years, cancer is now understood as a series of defects in the molecular machinery that governs proliferation and homeostasis in nearly all cell types. Normal cellular growth within an organism is kept in balance by various regulatory circuits that govern the rate at which cells divide, differentiate, and die. Some of these regulatory circuits are intrinsic to the cell whereas others are coupled to the signals that cells receive from their surrounding microenvironment. . . . Cancer arises through a process termed *neoplastic transformation* that occurs when a cell undergoes a series of genetic alterations and acquires the capability to escape these regulatory mechanisms. This process is thought to occur in a stepwise process involving the age-related incidence of four to seven stochastic events that drive the transformation of a normal cell into highly malignant clonal derivatives. . . . This process is similar to a Darwinian model of evolution, in that each genetic change confers a growth advantage that leads to overrepresentation of the altered cell. The successive and heritable nature of cellular transformation events is supported by histological analyses of precancerous lesions revealing cells that appear to represent intermediate steps in the pathway between normal and transformed cells. (Butterfield, Schoenberger, and Lyczak 2004 575)

> Most systematists agree that the animal kingdom is monophyletic; that is, if we could trace all animal lineages back to their origin, they would converge on a common ancestor. That ancestor was most likely a colonial flagellated protist that lived over 700 million years ago in the Precambrian era. This protist was probably related to choanoflagellates, a group that arose about a billion years ago. (Campbell and Reece 2002 634)

> The inflationary model for the early universe proposes that, starting about 10^{-34} seconds after expansion began, the rate of expansion began

> to increase rapidly with time. . . . during the next 10^{-34} seconds the universe doubled in size and continued to double in size during each succeeding 10^{-34} seconds until the inflationary epoch ended at a time of about 10^{-32} seconds. Because 10^{-32} seconds contains one hundred intervals each 10^{-34} seconds long, there was time for one hundred doublings of the size of the universe in the first 10^{-32} seconds. While inflation was going on, the universe grew in size by at least a factor of 10^{25} and perhaps much more depending on exactly how long inflation lasted. (Fix 2004 633)

Should we really believe that our best scientific theories simply tell us how things stand in the various inaccessible domains of nature they purport to describe? That is, are claims about polar covalent bonds, neoplastic transformations, and the first few moments of the universe's existence just like claims about the cherries in a cherry pie, aside from the fact that they concern things that are tiny, far away, rare, in the distant past, or otherwise hard to get information about? Or should we think of them in some other way, perhaps simply as useful conceptual tools for predicting natural phenomena and intervening to produce or prevent them, but not literal descriptions of how things stand in inaccessible domains of nature? For most of modern history (indeed, until the comparatively recent strict professionalization of different academic fields) this has been a central concern for scientists themselves as well as for others who thought seriously about the scientific enterprise. This was, for instance, the matter at the heart of the intense scientific debates throughout the eighteenth and nineteenth centuries (involving such luminary figures as Newton, Whewell, and Mill) concerning whether the "method of hypothesis" could produce genuine knowledge of nature. This same question has attracted considerable attention from a diverse array of thinkers in our own day. Although today we typically describe the dispute as concerned with something we call scientific realism, at the heart of the matter is still the simple question of whether or not we should understand our best scientific theories as literally true descriptions of how things stand in nature itself. Put perhaps more simply still, the issue is whether we should really *believe* what our best scientific theories say about the otherwise inaccessible parts of the world they seek to describe.

Of course, when we consider the fantastic practical achievements of contemporary science, it seems perverse if not insane to question whether our scientific theories simply tell it like it is. In introductory philosophy of science courses I sometimes break open a cheap shower radio to serve as a reminder of just how incredible even the most mundane of these technological accomplishments really are: a shower radio simply consists of little pieces of metal, glass, and plastic hooked together in a very specific way, and it strains credulity to think that such a device would allow you to shake your booty to K. C. and the Sunshine

Band's aptly titled "Shake Your Booty" (in the shower, no less) unless the theories we used to build it, concerning radio waves, electricity, acoustics, and much else besides, are accurate descriptions of how things stand in inaccessible domains of nature. More generally, the beliefs about the world delivered by theoretical natural science have afforded us powers of prediction, explanation, and intervention absolutely unrivaled in human history, and our usual, perfectly sensible, practice is to assume that beliefs able to successfully guide our practical engagement with the world in this way do so because they are *true*. In light of the spectacular and practical achievements of our best scientific theories, it might seem that the beliefs comprising those theories have a stronger claim on our credence than virtually any others we could name. And indeed, it seems fair to describe educated common sense about science as the view that the central and fundamental claims of our scientific theories must be at least *roughly* accurate, tempered by the awareness that we are surely in error about some matters of detail and that much of importance remains to be discovered.

This line of thinking has enjoyed widespread appeal not only among people of good common sense, but among professional philosophers of science as well, leading many of them to embrace a view of the matter entitled *scientific realism*: the position that the central claims of our best scientific theories about how things stand in nature must be at least probably and/or approximately true. The philosopher of science Hilary Putnam once formulated this line of argument in an especially memorable way, claiming that "[t]he positive argument for realism is that it is the only philosophy that doesn't make the success of science a miracle" (1975 73). The central idea of this influential "miracle argument" is that the only satisfactory explanation for the success of our scientific theories is that they are (at least approximately) true in the most straightforward sense of the term: any other view of the matter leaves it a complete and utter miracle why our best scientific theories are so successful. Powerful formulations and further developments of this *explanationist defense of realism* have been offered in recent decades not only by Putnam (1975, 1978); but also by Popper (1963); Smart (1968); Boyd (1984); Musgrave (1988); Leplin (1997); Psillos (1999); Kitcher (1993, 2001a, 2001b); and others. Because of its evident power and undeniable appeal, the explanationist defense is sometimes called the "ultimate argument" for scientific realism (first by van Fraassen 1980), I believe), and it has always been the strongest consideration in support of the realist position.

Despite the explanationist defense's deservedly wide influence, a few competing considerations have persistently encouraged a minority tradition among scientists and philosophers of science alike to question this apparently unassailable rationale for the realist position. Perhaps the most influential of these considerations concerns the historical record of scientific inquiry itself. We are not the first humans to develop theories about the natural world, nor are we the first to enjoy considerable and even incredible successes in prediction, intervention, or explanation by using those theories. Indeed, the history of science seems to consist of a succession of past theories that made radically different claims than our own theories do about the fundamental constitution and workings of nature, claims that were ultimately discovered to be false despite grounding just the same kinds of explanatory and predictive accomplishments that so impress us in contemporary theories. The challenge thereby posed to scientific realism was forcefully articulated by the French physicist Henri Poincaré near the turn of the twentieth century:

> The ephemeral nature of scientific theories takes by surprise the man of the world. Their brief period of prosperity ended, he sees them abandoned one after the other; he sees ruins piled upon ruins; he predicts that the theories in fashion today will in a short time succumb in their turn, and he concludes that they are absolutely in vain. This is what he calls the *bankruptcy of science*. ([1905] 1952 160)[1]

Contemporary philosophers of science call this argument the *pessimistic induction*: its central idea is that the scientific theories of the past have turned out to be false despite exhibiting just the same impressive sorts of virtues that present theories do, so we should expect our own successful theories to ultimately suffer the same fate. If the history of science really consists of a succession of increasingly successful theories making radically and fundamentally different claims about what there is in the world and how it works, why on earth would we suppose that this process has come to an end with the theories of the present day?

More recently, Larry Laudan (1981, 1984b) has used a similar appeal to the historical record of scientific inquiry to try simply to undermine the explanationist defense of realism itself. He argues that the historical record testifies that innumerable past scientific theories have been remarkably successful in just the same ways that our own theories are without being true, and that this in turn shows why the realist's inference from the success of a scientific theory to its truth is unwarranted. To feel the power of this argument, notice that defenders of past scientific theories occupied at one time just the same position that we do now: they thought the evident success in prediction, explanation, and intervention afforded us by, say, Newtonian mechanics rendered it impossible or extremely unlikely that the theory was false. If Newtonian mechanics weren't true, they might have said, then it would have to be a miracle that the theory is so successful and offers such accurate predictions and convincing systematic explanations concerning diverse physical phenomena ranging from the flight of cannonballs to the orbits of the planets. But they were wrong, and Laudan suggests that we would be equally wrong to draw the same conclusion about the successful theories of our own day: he argues that the successes of our own scientific theories do not constitute good evidence for the view that they simply state the facts about inaccessible domains of nature, because the history of science reveals that their rejected predecessors, Poincaré's "ruins piled upon ruins," turned out not to do so despite enjoying just the same sorts of predictive and explanatory achievements.

The other central consideration used to challenge the explanationist defense of scientific realism is called the *underdetermination of theories by the evidence*, and it is essentially concerned with the possible existence of alternatives to our best scientific theories that share some or all of their empirical implications—that is, quite different accounts of the entities and/or processes inhabiting some inaccessible domain of nature that nonetheless make the same confirmed predictions about what we should expect to find in the world and recommend the same successful strategies for intervening in it that our own theories do. No matter how impressive a theory's practical achievements in guiding prediction and intervention are, those achievements do not favor the theory over any alternative that would ground those same predictions and interventions and therefore enjoy just the same degree of empirical successes. And some philosophers of science have sought to show that *every* scientific theory must have what they call *empirical equivalents*: that is, alternatives sharing *all and only* the same empirical implications, which therefore supposedly cannot be better or worse confirmed by any possible body of empirical evidence. If this is so, it seems that we would be rash to believe our own theories to be true on the strength of the evidence we have, for we know that there are alternatives with just the same empirical credentials. The significance of these considerations has been defended perhaps most influentially by Bas van Fraassen (1980): his "constructive empiricism" argues that the threat of underdetermination should lead us to remain agnostic about the claims of even our best theories concerning entities or states of affairs that are unobservable, and to believe only what such theories claim or imply concerning observable matters of fact, for it is these latter claims that must be shared, of course, by any empirical equivalents that our theories may have.

The concerns about the truth of our best scientific theories prompted by the pessimistic induction and the problem of underdetermination are likely to strike many thoughtful people as perhaps interesting curiosities or intellectual puzzles but certainly nothing more—after all, how could it really be like that? Surely our best scientific theories couldn't really be so successful in their practical applications without being at least approximately true in their fundamental claims about nature. But in fact the history of science itself provides an abundant preserve of examples illustrating just what it is like for scientific theories to enjoy substantial empirical successes while being profoundly mistaken in their most fundamental claims about nature. According to our best contemporary physical theories, for example, gravity is not a force exerted by massive objects on one another, but instead reflects the curvature of space-time itself: gravitational motion is not like two marbles being pulled toward each other by invisible strings, but is instead like two marbles rolling from the lip to the bottom of a shallow bowl, where the bowl represents the deformation of the fabric of space and time itself produced by the masses of the marbles. Nonetheless, for a wide variety of purposes and in a wide variety of contexts, it is extremely useful to think of the world *as if* Newtonian mechanics were true—as if gravity were a force exerted by massive objects on one another, for instance. Indeed, Newtonian mechanics is still the physics we use to send rockets to the moon, because it is much simpler to work with than the contemporary alternatives and the empirical predictions it makes at the scale of rockets and moons turn out to be quite accurate despite the fact that it is profoundly mistaken about the fundamental constitution of nature. Although this sometimes invites the counterclaim that Newtonian mechanics itself is "approximately true," this can only mean that its empirical predictions approximate those of its successors across a wide range of contexts. It cannot mean that it is approximately correct as a fundamental description of the physical world: in this respect, Newtonian mechanics is just plain false, and radically so. And this recognition invites us to ask, though perhaps in a whisper, whether it might not be that all of our own scientific theories are both fundamentally mistaken and nonetheless empirically successful *in just this same way*?

This is undoubtedly a heady possibility, but it is one that we would be rash to embrace too hastily. In the next section we will look at some reasons that have been given for doubting that the pessimistic induction and the underdetermination of theories by evidence really should lead us to withhold belief in the claims of our best scientific theories. As it turns out, I will suggest, the most powerful challenge to scientific realism has yet to be formulated but emerges naturally from a systematic consideration of the reasons that the much more famous challenges offered by the pessimistic induction and the underdetermination of theories by evidence have left scientific realists unconvinced of their significance.

1.2 Problems for Pessimism and Underdetermination

Part of the appeal of the classical pessimistic induction is that the argument it offers is so very straightforward. It depends on no controversial theories or contentious claims about the nature of scientific inquiry. Instead, it simply points out that past scientific theories have repeatedly turned out to be radically false despite enjoying just the same sorts of predictive and explanatory successes that our own theories do, and concludes from this evidence that the same fate probably awaits our own contemporary scientific theories as well. But the very simplicity that makes this challenge so striking also invites a natural reply to it, for the argument itself relies on an extremely simple form of reasoning called enumerative induction (also known as inductive generalization or the "straight rule" of induction) that is vulnerable to an obvious sort of objection. That is, the pessimistic induction simply projects in a straightforward fashion from what has happened in past cases to what will happen in present and future ones, and it is an easy matter to point out cases in which reasoning in this simplistic way would produce mistaken or even contradictory results. I might, for instance, infer my immortality from the fact that I have (eventually) risen to greet the sunshine every day since I was born, or my eventual demise from the deaths (before a certain age) of all those who have come before me. Such enumerative induction is a relatively clumsy inferential tool that can lead us astray in any number of ways, particularly when circumstances change in some way that is relevant to the continuity of the regularity or mechanism that grounds it.

And of course we already know that there are indeed many important differences, both in general and in particular cases, between present scientific theories and those ultimately rejected predecessors that would have to form the evidential basis for any inductive generalization we might try to extend to them. It seems perfectly natural to say, for instance, that past theories enjoyed many of the same virtues and varieties of success as even the best of their present counterparts, but not to the same degree. Accordingly, it is common to find scientific realists objecting that the pessimistic induction seeks to generalize unfairly from theories found in the 'immature' periods of various sciences and/or those not matching the performance of one or more contemporary theories in some particular respect. That is, realists can point to the breadth of application some present theories have achieved, the diversity and precision of the empirical predictions they offer, their success in predicting novel phenomena not known before and/or used to develop the theory itself or other such features, and suggest with some justice that these characteristics distinguish at least some present theories from at least some of their ultimately rejected predecessors in a way that might invalidate the pessimistic induction's projection from the fate of the latter to the prospects for the former. The lingering whiff of ad-hoc-ery or special pleading cannot dispel the fact that changed circumstances, conditions, or characteristics sometimes really do make a difference to the legitimacy of projecting an inductive generalization into the future.

In this way, disputes over the legitimacy of the pessimistic induction seem to have reached something of a stalemate in the philosophy of science. Realists respond to the challenge of the pessimistic induction by pointing out ways in which at least some contemporary theories are indeed distinct from their predecessors and therefore reject the validity of the inductive projection from past to present cases. In response, defenders of the pessimistic induction demand to know why just *these* varieties or degrees of success are special in a way that should block the proposed inductive projection, when earlier varieties and degrees of predictive and explanatory success that were equally thought to be explicable only by the truth of the theories that enjoyed them turned out not to be so.[2] In the resulting standoff, each side simply seems to shift the burden of proof onto the other in a way that is neither altogether illegitimate nor altogether convincing: surely some differences of this sort would invalidate the inductive projection from the fortunes of past theories to the prospects for present ones, but whether these particular differences should insulate present theories from historical comparison in this way would seem to be anyone's guess.

There is a similarly stalemated character to the arguments presently offered on both sides of the challenge posed to scientific realism by the underdetermination of theories by the evidence. As we noted above, the problem of underdetermination grows out of the worry that there might be alternatives to even our best scientific theories that would make the same predictions and recommend the same interventions in the cases we have tested, and would therefore be no less well confirmed than our own theories by the dramatic empirical successes those theories have achieved. But it is quite difficult to decide how seriously we ought to take this frankly speculative possibility. In the absence of any evidence, why should we either assume that such alternatives exist or let the bare possibility that they might exist prevent us from believing the best-confirmed theories we do have? It seems perfectly sensible for critics of underdetermination to insist that any such equally well-confirmed alternatives *actually be produced* before we take them seriously or suspend judgment about the truth of the contemporary theories whose achievements they are supposedly able to replicate (see Kitcher 1993, Leplin 1997, Achinstein 2002).

Faced with this curmudgeonly lack of enthusiasm for such an obviously exciting prospect as the underdetermination of theories by evidence, philosophers of science sympathetic to the idea have accepted what I will now suggest is something of a devil's bargain in their efforts to show that it is anything more than a speculative possibility. They have concentrated their attention and argumentative efforts on the rather trivial forms of underdetermination that they can *prove* to obtain universally, and in the process they have unwittingly abandoned the effort to show that underdetermination obtains generally in any sense that should actually lead us to question the truth of our best scientific theories. More specifically, defenders of underdetermination have sought to provide us with a procedure (ideally an algorithmic or mechanical procedure) for generating empirical equivalents to any theory at all, irrespective of its content, character, or subject matter. That is, they have sought to articulate a procedure for generating alternatives to absolutely any theory that will have precisely the same empirical implications as the original and will therefore supposedly be indistinguishable from it by any possible evidence. It is this strategy of trying to defend the significance of underdetermination by showing that all theories either do or must have empirical equivalents which I suggest constitutes a devil's bargain for defenders of underdetermination, for it succeeds only where it gives up any significant and distinctive general challenge to the truth of our best scientific theories. I will discuss the failings of this strategy in considerable detail, because seeing just how the search for empirical equivalents fails to establish any significant general problem of underdetermination will give us our first indication of how a much more serious challenge to scientific realism can be developed.

Algorithms for generating empirical equivalents fall roughly but reliably into *global* and *local* varieties. Global algorithms are designed to produce empirical equivalents from absolutely any theory and are perhaps best exemplified by André Kukla's appeals to such all-purpose alternatives to any theory T as the following:

1. T'—the claim that T's observable consequences are true, but T itself is false
2. T''—the claim that the world behaves according to T when observed, but some specific incompatible alternative otherwise
3. The hypothesis of the Makers—the debatably coherent fantasy that we and our apparently T-governed world are part of an elaborate computer simulation
4. The hypothesis of the Manipulators—the claim that our experience is manipulated by powerful beings in such a way as to make it appear that T is true. (1996; see also Kukla 1993 and van Fraassen 1985)

Kukla devotes his efforts to defending such proposals from the accusation (see Laudan and Leplin 1991, Hoefer and Rosenberg 1994) that they are not "real theories" at all. But this is beside the point, I suggest, for *whether or not* such farfetched scenarios are real theories they amount to no more than a salient reminder in a scientific context of the general possibility of the sort of radical skepticism captured by a famous thought experiment developed by Descartes: that there might be an all-powerful "Evil Demon" who devotes his energies to deceiving us about what the world is really like. Such philosophical fantasies are the engine of traditional radical or Cartesian skepticism: they offer an equally powerful (or powerless) challenge to absolutely *any* knowledge claim whatsoever, however derived or supported, for the powers of the Evil Demon know no limits. While many contemporary philosophers are inclined to grant that such radical skepticism cannot be refuted on its own terms, underdetermination was supposed to represent a distinctive epistemic problem arising specifically or at least perspicuously in the context of scientific theorizing about inaccessible domains of nature and thus troubling even those sensible souls who never hoped to defend their scientific beliefs to the truly radical skeptic in the first place. Thus, if

Cartesian fantasies are the only reasons we can give for taking the possibility of underdetermination seriously, then there simply *is no* distinctive problem of scientific underdetermination to worry about, for the worry *just is* the specter of radical skepticism familiar from introductory philosophy courses everywhere.[3] Perhaps we need an answer (or perhaps there is no answer) to Descartes' Evil Demon, but there is no problem or challenge with special significance for theoretical science to be found here.

The same response applies to some famous nonalgorithmic examples of empirical equivalents, like the notorious prospect of a continuously shrinking universe with compensatory changes in physical constants making this state of affairs undetectable to us (that is, theories we describe as making unmotivated and/or wildly implausible assumptions about nature). Some judgments of prior plausibility are required in order to escape radical or Cartesian skepticism in the first place, and we are no less entitled to these resources in a scientific context than any other.[4] Whether general or specific, Cartesian fantasies again simply *replace* our worry about scientific underdetermination with a quite different (perhaps insoluble, but familiar) general skeptical problem.

A similarly subtle change of subject arises with the demand that we consider what is sometimes called the 'Craigian reduction' of a theory (that is, a statement of that theory's observable consequences) as a competitor when trying to assess the plausible threat of underdetermination. Perhaps even Craigian reductions are "real theories," but the underdetermination worry was supposed to be that there might be too many different theoretical accounts of the inaccessible workings of nature well confirmed by the evidence, *not* simply that there are (as we already knew) multiple options for beliefs about the world that the evidence leaves us free to accept. Agnosticism about all accounts of the inaccessible aspects of nature is always possible, but is *defensible* only if the underdetermination of theory by evidence (or some other ground for suspicion about all theories) is independently established, for we surely want the strongest set of beliefs to which we are entitled by the evidence. It is not enough that the epistemically more modest choice to believe only a theory's claims about observable phenomena is always left open by the evidence (cf. van Fraassen 1980); for that matter, so is choosing to believe nothing at all.

By contrast to the global strategy's Cartesian fantasies, the local algorithmic strategy seeks instead to take advantage of one or more formal features of a particular theory to show that an infinite or indefinite number of serious scientific empirical equivalents to that theory can be produced by varying the feature(s) in question. Consider the now-famous example of TN(0): Newtonian mechanics and gravitational theory, including Newton's claim that the universe is at rest in absolute space. This theory supports any number of empirical equivalents of the form TN(v), where v ascribes some constant absolute velocity to the universe.

But such empirical equivalents prove too little. The sensible realist will surely insist that we are not here faced with a range of competing theories making identical predictions about the observable phenomena, but instead just a single theory being conjoined to various factual claims about the world for which *that very theory* (along with other beliefs we hold) implies that we cannot have any empirical evidence.[5] It is by no means always trivial to determine which elements of a proposed theory are otiose by its own lights, but the sensible realist will counsel realism only about those theoretical claims (*whatever* they are) that our theories themselves imply are amenable to empirical investigation. This realism should no more extend to the conjunction of Newtonian theory with claims about the absolute velocity of the universe than with claims about the existence of God.

Another way to see this point is to note that empirical equivalents of the TN(v) variety pose no threat to the *approximate* truth of our theories: if the realist believes TN(0) when one of the various TN(v) obtains, most of her theoretical beliefs about the relevant domain will be straightforwardly true. Thus, empirical equivalents of the TN(v) variety show at most that we would have been unjustified in taking any stand on the constant absolute velocity of the universe, not in accepting the other theoretical claims of Newton's theory.

Our response to the local algorithmic strategy, like the global, applies equally well to some famous nonalgorithmic examples. John Earman suggests (drawing on results from Clark Glymour and David Malament), for example, that underdetermination threatens because "even idealized observers who live forever may be unable to empirically distinguish hypotheses about global topological features of some of the cosmological models allowed by Einstein's field equations for gravitation" (1993 31).[6] But such claims about global topology—concerning, for example, the compactness of space (as determined relative to some canonical foliation of spacetime)—are simply factual claims about the world for which the general theory of relativity itself (again, given accepted auxiliary hypotheses) suggests that we are (or may be) unable to acquire evidence. And there is surely something pathological about the claim that hooking one or another such claim to the general theory of relativity produces genuinely distinct, empirically indistinguishable theories: once again, the sensible realist will surely counsel realism only about those aspects of well-confirmed theories that *those theories themselves* hold to be empirically significant.

This suggests that the local strategy (like the global) actually trades in underdetermination for another long-standing philosophical problem, this time a puzzle in the theory of confirmation sometimes called the "tacking" problem: if the true empirical consequences of a theory are all that matters to its confirmation, then evidence E confirming theory T will equally well confirm theory T + C (where C is any further claim that does not undermine T's implication of E), thus offering spurious confirmation to C itself. That is, suppose we "tack on" to contemporary molecular genetics the further claim "and jellybeans grow on trees on the planets orbiting Alpha Centauri": if true empirical consequences are all that matter to confirmation, this new "theory" (including its claim about the jellybean trees) will be just as well confirmed as molecular genetics itself, for it shares all the successful empirical predictions of the latter. This case is ridiculous, of course, but notice that this is precisely (though less obviously) what happens in cases like the Newtonian claim that the universe is at absolute rest and in the relativistic claims about global topology discussed above. Like Cartesian skepticism, this tacking problem is philosophically serious, indeed it requires some solution, but it cannot be our only reason for taking *underdetermination* seriously without simply collapsing the latter problem into the tacking problem itself. If it is, then we once again have no distinctive problem of underdetermination to worry about, only a cleverly disguised illustration of why any successful account of genuine confirmation will have to be able to exclude confirmation of this pathological "jellybean tree" variety.

In retrospect, perhaps it should not surprise us that philosophers' algorithms cannot make short work of the daunting task of generating alternative hypotheses that are both scientifically serious and genuinely distinct from existing competitors, for this is precisely the sort of difficult conceptual achievement that demands the sustained efforts of real scientists over years, decades, and even careers. That is, perhaps the very attempt to develop a formal, algorithmic procedure for demonstrating the existence of the sort of empirical equivalents that would actually give us grounds for concern about the truth of our own scientific theories should always have struck us as far too ambitious in precisely the wrong way.

There are, of course, particular examples of empirical equivalents that are neither skeptical fantasies nor trivial variations on a single theory. Perhaps most convincing is Earman's other supporting case: "TN (*sans* absolute space)...opposed by a theory which eschews gravitational force in favor of a nonflat affine connection and which predicts exactly the same particle orbits as TN for gravitationally interacting particles" (1993 31; see also Glymour 1977). Neither theory is a skeptical fantasy, nor is one a trivial variation on the other: treating gravitational attraction as a fundamental force seems substantially different from treating it as manifesting the curvature of spacetime. Other plausible (albeit more controversial) cases include special relativity versus Lorentzian mechanics (controversial because the latter's requirement of systematic expansion and contraction for all our measuring devices (including rods and clocks irrespective of internal composition or construction) when in motion relative to absolute space might be thought a skeptical fantasy) and Bohmian hidden variable versus standard Von Neumann-Dirac formulations of quantum mechanics (controversial because it is not clear that we understand quantum mechanics well enough to say convincingly what formulations of it count as genuinely different theories).[7]

Of course, the convincing examples are drawn exclusively from the physical sciences and, as Laudan and Leplin rightly point out (1991 459), typically involve the relativity of motion in one guise or another. This idiosyncrasy might lead us to suspect that they form an unrepresentative sample and/or that there is something about the characteristic structure of physical theories (if such there be) that renders them especially susceptible to the construction of empirical equivalents: biologists and philosophers of biology, for example, have no idea how they would go about constructing even one (genuinely distinct, nonskeptical) empirically equivalent alternative to the modern synthesis of Darwinian evolutionary theory and Mendelian genetics. Much more importantly, *none* of these examples is generated by an algorithm or formula and each is a hard-won *particular* alternative to an existing theory (rather than an infinite or indefinite collection) that proved quite difficult to identify and characterize: surely one or even a few such convincing cases do not provide sufficient warrant for concluding that genuine or serious empirical equivalence is a ubiquitous phenomenon. If numerous serious empirical equivalents to virtually any theory could be produced with just a little determined effort and ingenuity, this would certainly ground the worry that an infinite space of equally well-confirmed alternatives looms over each of our scientific theories, but the profound difficulties and rare success we have encountered in trying to develop even one or a few convincing examples of nonskeptical and genuinely distinct empirical equivalents might sensibly be seen to support just the opposite general conclusion.

Thus, the case for underdetermination from empirical equivalents will simply not support the intoxicating morals that advocates hoped to draw. Algorithms provide proofs of the underdetermination predicament only by transforming the problem into one venerable philosophical chestnut or another, while one or a few convincing examples, dearly purchased and drawn from a single domain of scientific theorizing, are unable to support the sweeping conclusion that there are likely serious empirical equivalents to most theories in most domains of scientific inquiry. Scientists and philosophers concerned with a particular theory would surely do well to worry about whether *that theory* has genuine empirical equivalents. But it seems that the critics of underdetermination have been well within their rights to demand that serious, nonskeptical, and genuinely distinct empirical equivalents to a theory actually be produced before they withhold belief in it and refusing to presume that such equivalents exist when none can be identified.

1.3 Recurrent, Transient Underdetermination, and a New Induction over the History of Science

Of course, the search for empirical equivalents was only the most promising strategy for trying to *prove* that underdetermination *always* obtains. It is therefore somewhat alarming that the connection between the two issues has become so firmly established that the most influential (and ostensibly general) recent attack on underdetermination (Laudan and Leplin 1991) and its most influential (and ostensibly general) recent defense (Earman 1993) both proceed *solely* by addressing the existence and status of putative empirical equivalents.[8] But the lack of any convincing case for the widespread existence of significant empirical equivalents simply does not settle the seriousness with which we should regard the threat of underdetermination itself. That threat was not initially concerned with the possibility of empirical equivalents at all, of course, but instead with *any* alternatives sharing the impressive empirical achievements of our own best scientific theories. Notice, for instance, that our grounds for belief in a given theory would be no less severely challenged if we believed that there are one or more alternatives that are not empirically equivalent to it but are nonetheless consistent with or even equally well confirmed by *all of the actual evidence we happen to have in hand at the moment*. Following Larry Sklar (1975), we might call this a *transient* underdetermination predicament: that is, one in which the theories underdetermined by the existing evidence are empirically inequivalent and could therefore be differentially confirmed by the accumulation of further evidence. Little-noticed in the heated crossfire over empirical equivalents is the fact that even such a transient underdetermination predicament undermines our justification for believing present theories in general, so long as we have some reason to think that it is also *recurrent*: that is, that there is (probably) at least one such alternative available (and thus this transient predicament rearises) *whenever* we must decide whether to believe a given theory on the strength of a given body of evidence.

Of course, it seems clear that we do not occupy such a predicament of recurrent, transient underdetermination if we consider only the theoretical alternatives we have in fact developed and considered to date: as a general matter, we think our own scientific theories are considerably better confirmed by the evidence we have in hand than any of the competing accounts of nature we have actually produced to this point in the history of scientific inquiry. Thus, the danger of recurrent, transient underdetermination does not even threaten to become acute unless we consider the possibility that there might be such empirically inequivalent but nonetheless well-confirmed, serious alternatives among the theories that we have not yet even imagined or entertained. I am suggesting, that is, that any real threat from the problem of underdetermination comes not from the sorts of philosophically inspired theoretical alternatives that we can construct parasitically so as to perfectly mimic the predictive and explanatory achievements of our own theories, but instead from ordinary theoretical alternatives of the garden variety scientific sort that we have nonetheless simply not yet managed to conceive of in the first place. I will call this worry the *problem of unconceived alternatives*, and although it has historically received far less attention than either the search for empirical equivalents or the traditional pessimistic induction, I will suggest that it ultimately deserves far more.

Again the tough question, of course, is how to decide whether or not there really *are* typically unconceived competitors to our best scientific theories that are well confirmed by the body of actual evidence we have in hand. To decide this we will need to know something about the set of hypotheses we *haven't yet considered*; specifically, whether it includes scientifically plausible competitors to our best scientific theories offering equally convincing explanations of the phenomena and therefore having an equally strong claim to represent the theoretical truth about nature. And of course, it is not easy to acquire compelling evidence about the existence of hypotheses that are, *ex hypothesi*, unconceived by us. Sklar (1981) represents the most notable exception to the general neglect of the threat of recurrent, transient underdetermination, but he finds it reasonable to simply *assume* in light of "the limitations of our scientific imagination" (and, at least in part, "reflection upon historical scientific experience," a suggestion I will try to flesh out in the remainder of this chapter) "that there are vast numbers of perfectly respectable scientific hypotheses... we just haven't yet brought to mind" including "innumerable alternatives to our best present theories... which would save the current data equally well" and probably even some "more plausible than our own theories relative to present observational facts" (1981 18–19). Elsewhere (1975 381) he simply supposes without defense that even those who are skeptical of empirical equivalence "are likely to admit that transient underdetermination is a fact of epistemic life."[9] But as we've already seen, *these are just the claims with which critics of underdetermination will take issue on any nontrivial reading* (cf. Kitcher 1993, Leplin 1997, Achinstein 2002).

While it is obviously difficult to acquire convincing evidence regarding the likely existence and character of presently unconceived theories, I think that there is a genuine argument to be made out of Sklar's brief but tantalizing intimation that the history of science itself bears on this question. Indeed, I suggest that the historical record of scientific inquiry provides compelling evidence that recurrent, transient underdetermination is our actual epistemic predicament in theoretical science rather than a speculative possibility. I would also suggest, however, that this very historical record contradicts Sklar's further suggestion (1981 22ff) that the threat can be substantiated only for fundamental physical or cosmological theories, as well as his more recent efforts to soften the sting of this worry with the suggestions that the historical progression of such theories is largely "one in which each successor theory is framed in concepts that are refinements and deepenings of the concepts of the theory that preceded it" (2000 94) and that well-confirmed past theories can be seen as having been "pointing towards" later alternatives or "heading in the right direction" (2000 chap. 4, section I, passim). Instead, I suggest that the historical record offers plainspoken inductive testimony to the fact that we have repeatedly occupied a predicament of recurrent, transient underdetermination across a wide and heterogeneous variety of scientific fields and domains of inquiry simply because we have repeatedly failed to conceive of all the empirically inequivalent but scientifically serious alternative theoretical possibilities well confirmed by the evidence available to us.

Recall that the classical pessimistic induction notes simply that past successful theories have turned out to be false and suggests that we have no reason to think that present successful theories will not suffer the same fate. By contrast, I propose what I will call the *new induction over the history of science*: that we have, throughout the history of scientific inquiry and in virtually every scientific field, repeatedly occupied an epistemic position in which we could conceive of only one or a few theories that were well confirmed by the available evidence, while subsequent inquiry would routinely (if not invariably) reveal further, radically distinct alternatives as well confirmed by the previously available evidence as those we were inclined to accept on the strength of that evidence.[10] For example, in the historical progression from Aristotelian to Cartesian to Newtonian to contemporary mechanical theories, the evidence available at the time each earlier theory was accepted offered equally strong support to each of the (then-unimagined) later alternatives. To be sure, the theory of relativity might never have been developed were it not for the evidential anomalies that emerged for Newtonian mechanics, but the radically different theoretical account of gravitational motion offered by the former was nonetheless equally well supported by the many phenomena for which the latter already provided a convincing account. In a similar fashion, I suggest, we have repeatedly found ourselves encouraged or even forced under the impetus provided by recalcitrant phenomena, unexpected anomalies, and other theoretical pressures to discover new theories that had remained previously unconceived despite being well confirmed by the evidence available to us. This same pattern would seem to characterize any number of similarly important theoretical progressions and/or transitions in the history of science, including such famous examples as the following:

> from elemental to early corpuscularian chemistry to Stahl's phlogiston theory to Lavoisier's oxygen chemistry to Daltonian atomic and contemporary chemistry

from various versions of preformationism to epigenetic theories of embryology

from the caloric theory of heat to later and ultimately contemporary thermodynamic theories

from effluvial theories of electricity and magnetism to theories of the electromagnetic ether and contemporary electromagnetism

from humoral imbalance to miasmatic to contagion and ultimately germ theories of disease

from eighteenth century corpuscular theories of light to nineteenth century wave theories to the contemporary quantum mechanical conception

from Darwin's pangenesis theory of inheritance to Weismann's germ-plasm theory to Mendelian and then contemporary molecular genetics

from Cuvier's theory of functionally integrated and necessarily static biological species and from Lamarck's autogenesis to Darwin's evolutionary theory

These prominent examples at least suggest a robust, distinctive pattern, in which the available evidence cited in support of each earlier theory ultimately turned out to support one or more competitors unimagined at the time just as well. Many less famous examples of theoretical progressions could presumably be added as well, but even this fairly short list suffices to illustrate that the pattern is characteristic of theoretical science across a wide variety of fields and historical circumstances.[11] Thus, the history of scientific inquiry itself offers a straightforward rationale for thinking that there typically are alternatives to our best theories equally well-confirmed by the evidence, even when we are unable to conceive of them at the time.

In at least some of these cases, however, it will surely be objected that changes in accepted background scientific beliefs or auxiliary hypotheses were required before the alternatives in question could rightly be regarded as well confirmed by the available evidence. This is so, but it ignores the new induction's suggestion that in such cases the needed alternative auxiliary hypotheses will often or even typically *themselves* be ones that were unconceived despite being supported in an equally compelling fashion by the available evidence. In other words, the new induction suggests that in such cases the totality of evidence available at the time of an earlier theory's acceptance characteristically offers equally compelling support for the combination of a later accepted alternative to that theory together with the requisite alternative auxiliary hypotheses that would themselves later come to be accepted. And such a combination must surely be regarded as a scientifically serious alternative possibility rather than a mere skeptical fantasy, for it is ultimately accepted by some actual scientific community.

This should help to clarify why the central claim of this new induction does not rest on an atavistic account of confirmation on which the evidential relation is construed merely as subsumption or on which the only constraint on evidential warrant is the conformity of evidence to the theory: indeed, it is intended to be plausible no matter what account of confirmation we favor. We cannot, for example, mitigate the problem by assigning widely divergent prior probabilities to earlier and later theories, for all the theories under consideration (unlike the radical skeptic's fantasies) are demonstrably serious scientific possibilities in the only sense that matters here: after all, each is ultimately accepted by some actual scientific community. Nor will it help to appeal to the differential confirmation offered to an hypothesis by its fit with other theories well confirmed and/or accepted at the time (cf. Boyd 1973, Laudan and Leplin 1991) if the new induction is right to suggest (above) that the later alternatives to such further theories are often or even typically themselves unconceived despite also being supported by the available evidence. Indeed, while Laudan and Leplin have themselves offered a convincing general attack on the conflation of evidential warrant with mere subsumption or verification of a theory's strict empirical entailments (and of the conflation of epistemic notions and issues with syntactic ones quite broadly), the intuitive and robustly epistemic conception of confirmation they offer as a corrective is the

> general idea about theory testing and evaluation . . . that there [is] a range of 'phenomena' for which any theory in a particular field [is] epistemically accountable. . . . For a Newton, a Ptolemy, or a Mach, 'saving the phenomena' meant being able to explain all the salient facts in the relevant domain. (1991 471)

And of course, this is precisely the intuitive conception of evidential warrant on which the new induction's central claim, that the evidence used to support earlier theories turns out to provide equally compelling support for alternatives unconceived at the time, seems most plausible and convincing.

For this historical claim to be even remotely plausible, we must explicitly note that a theory need not explain or accommodate all the existing data in order to count as well confirmed: evidential anomalies are allowed. The point is that we have repeatedly been able to conceive of only a single theory that was well supported by all of the available evidence when there were indeed alternative possibilities *also* well supported, indeed perhaps equally or even better supported, by that same body of evidence. Nor, therefore, does this suggestion ignore the phenomenon of explanatory losses in the transition from an earlier theory to a later one. That is, a theory need not explain everything that a competitor explains in order to be as well supported by the totality of available evidence. The theories may simply have *different* accomplishments and/or evidential anomalies. Thus, it need not trouble us that Aristotelian mechanics was used to explain the generation of cats and the formation of human societies, while Newtonian mechanics had no such explanatory ambitions. On the other hand, the judgment that alternatives not yet conceived were at least roughly as well supported by the available evidence as earlier competitors will require us to simply reject the most radical claims of "incommensurability" influentially defended by Thomas Kuhn ([1962] 1996), on which the very phenomena themselves quite literally do not *exist* in any way that permits their identification across theories or theoretical paradigms. That is, defenders of the new induction must insist that the constrained motion of rocks in slings is a single phenomenon described differently by Aristotelian and Newtonian mechanics, rather than allowing that the first recognized only a mixture of natural and violent motion while the second recognized only the completely distinct phenomenon of pendula losing energy through friction.[12] But of course, such radical incommensurability has always been among the most contentious and least plausible features of Kuhn's view of science, and accepting it would seem to offer scant comfort to scientific realism in any case.[13]

Nonetheless, this new induction will disappoint many champions of underdetermination, for the historical record offers at best fallible evidence that we presently occupy a significant underdetermination predicament, rather than the sort of demonstrative proof that advocates have traditionally sought (and thus far I have been able to do no more than suggest that this is indeed the verdict of the historical record). Furthermore, unlike constructing empirical equivalents, it does not allow us to say just *which* actual theories are underdetermined by the evidence, nor anything about what the (unconceived) competitors to present theories look like. On the other hand, I have suggested that the search for empirical equivalents has managed to provide convincing evidence of an underdetermination predicament only where it has transformed the problem into one or another familiar philosophical puzzle. These forms of underdetermination simply do not threaten to bear out the original concern that the very same evidence leading us to embrace our own scientific theories might turn out to support alternative scientific accounts of the same inaccessible domains of nature just as well. Abandoning our lingering fascination with them in this connection seems a small price to pay for returning our attention to the sort of underdetermination that the historical record suggests might pose a substantial challenge to even our most impressive achievements in the distinctive epistemic context of scientific theorizing about nature.

Thus, the problem of unconceived alternatives concerns alternatives to our best scientific theories, but not in the same way that the search for empirical equivalents does. Furthermore, it draws its force and evidence of its significance from the historical record of scientific inquiry, but not in the same way that the traditional pessimistic induction does. At its heart is neither the simple historical revelation that even the best scientific theories of any given earlier day have turned out to be false, nor the concern that we might be able to generate alternative theories that cannot be better or worse supported than our own by any possible evidence. Instead, the problem of unconceived alternatives worries that there are theories that we should and/or would take seriously as competitors to our best accounts of nature if we knew about them, and that could or have been distinguished from them evidentially, but that are excluded from competition only because we have not conceived of or considered them at all.

In the next chapter we will seek to refine our grasp of the problem posed by such unconceived alternatives by exploring the precise character and scope of the challenge they raise for scientific realism. There I note that the problem poses a credible challenge only to a particular sort of scientific inference in a particular kind of epistemic context: our efforts to confirm or verify scientific hypotheses by eliminating competing alternatives until only a single well-supported candidate theory remains. But I will suggest that it is our vulnerability to the problem of unconceived alternatives in just this context that is most significant for the dispute over scientific realism and for which we have the strongest historical evidence. We will also return to explore some of the important connections and differences between this challenge for scientific realism and that posed by the original pessimistic induction.

Chapters 3 through 5 will seek to deliver on at least a few of the promises I have made concerning the historical evidence. I will first use some brief examples to illustrate the significance of the problem of unconceived alternatives in the early history of modern theorizing about problems of generation and inheritance (traditionally dated from the publication of William Harvey's *De Generatione Animalium* in 1651). I will then go on to offer a sustained defense of the claim that our vulnerability to this problem persisted even after the search for both a material substrate of heredity and a mechanical account of the process of transmission ushered in the era of theorizing about inheritance both methodologically and substantively continuous with our own. More specifically, I will argue that the theorists of inheritance and generation in this contemporary tradition, including Charles Darwin, Francis Galton, and August Weismann, repeatedly failed even to conceive of important and scientifically serious alternative lines of theoretical development offering equally convincing explanations of the empirical phenomena for which they sought to account.

Chapters 6 and 7 will take up a variety of recent realist replies to the pessimistic induction that, if successful, would threaten to dispose of the problem of unconceived alternatives as well. Hardin and Rosenberg, Kitcher, Psillos, Leplin, and Worrall have offered such replies, and these have traditionally taken one of two fundamental forms. First, some realists have argued that there is more substantive continuity between earlier, rejected theories and their successors than the pessimistic induction allows. If so, we might well worry that successful later theories were never really *unconceived* alternatives to their predecessors in the first place. Second, realists have sometimes argued that we can know when our accounts of nature are (at least approximately) true notwithstanding the long track record of fundamentally mistaken past scientific theories that enjoyed (at least some kinds of) scientific success. If this were so, we might well be in a position to responsibly judge that some current theories are true even if there are serious alternatives to them that remain unconceived. I will argue that both of these strategies manage to achieve only Pyrrhic victories for realism: in every case they concede either just the substantive points that the realist was concerned to defend or everything her opponent needs to build a further convincing historical case against realism itself. Thus, the most sophisticated recent efforts to respond to concerns about realism grounded in the historical record do not even manage a convincing reply to the original pessimistic induction to which they are addressed, much less to the problem of unconceived alternatives they fail to anticipate.

Chapter 8 will begin to address the question of whether there is any sense to be made of science without scientific realism. That is, it will try to identify whether there is any coherent positive view we might take of our successful scientific theories if we abandon the presumption that they must be approximately true descriptions of nature's innermost recesses and secret domains. A long and distinguished minority tradition has embraced a view entitled "instrumentalism," which instead regards even our best scientific theories merely as effective tools or instruments for achieving our practical goals. Traditionally, such an instrumentalist view of science has been associated with a variety of implausible claims about the very meaning of theoretical claims, but at its heart is the simple idea that conceptual resources like scientific theories can be useful guides to prediction, to action, and to further investigation *without* being literally or even approximately true, in just the way that we now think Newtonian mechanics was both radically false and nonetheless an extremely useful instrument. The problem of unconceived alternatives promises to breathe new life into this instrumentalist tradition: not into its discredited semantic theses, of course, but into its positive conception of the status of scientific theories. For if the problem of unconceived alternatives is as pervasive as I suggest and has the implications that I claim, the natural conclusion to draw will be that the fundamental theories of contemporary science should be regarded, like their historical predecessors, simply as powerful conceptual tools for action and guides to further inquiry rather than accurate descriptions of how things stand in otherwise inaccessible domains of nature.

Notes

1. We should perhaps note explicitly that Poincaré does not ultimately embrace such an abject pessimism about science; for more detailed discussion, see chapter 8.

2. In a running gag in the classic "Peanuts" comic strip, Lucy would hold a football and get Charlie Brown to come running up to kick it, only to have her snatch it out of the way at the last moment and let him land flat on his back. But of course each time she talked him into trying this once again, Lucy had a new reason to convince Charlie Brown that this time she would finally let him kick the ball.

3. Kukla sometimes appreciates the Cartesian character of his scenarios (e.g. 1996 158), but not how this undermines the significance of his case for underdetermination.

4. We cannot evade every worry about underdetermination by appeal to the need for judgments of prior plausibility or likelihood, however, for the prior plausibility of scientific alternatives like electrons, phlogiston, and curved spacetime is simply not on a par with that of Cartesian Evil Demons (cf. Van Fraassen 1980 36). I suspect that this difference is what is really at issue in the somewhat misleading claim that some scenarios are too farfetched to constitute "real theories" at all (e.g. Leplin and Laudan 1993 11).

5. Although drawing this implication from the theory may require assumptions beyond the claims of the theory itself, these will be just the *same* further assumptions needed to assert the empirical equivalence of the various TN(v). Of course, changes in the accepted auxiliary assumptions over time may defeat the claim of empirical equivalence (a central point in Laudan and Leplin's (1991) attack on underdetermination), but in this context we are concerned with what to make of the prospect of theories that are empirically equivalent *given* (or "indexed to." see Kukla 1993) a particular set of auxiliary assumptions, or alternatively, with empirically equivalent "global theories" or "systems of the world" (see Hoefer and Rosenberg 1994).

6. For Earman (1993), the crucial sense of empirical equivalence (his EI_3) obtains between two hypotheses just in case two worlds in which those two hypotheses are respectively true need not be distinguished by some piece of empirical evidence. The differences between various possible formulations of empirical equivalence will not matter for the points at issue here.

7. While Eddington, Reichenbach, Schlick and others have famously agreed that general relativity is empirically equivalent to a Newtonian gravitational theory with compensating "universal forces," the Newtonian variant has never been given a precise mathematical formulation (the talk of universal forces is invariably left as a promissory note), and it is not at all clear that it can be given one. (David Malament has made this point to me in conversation.) The "forces" in question would have to act in ways no ordinary forces act (including gravitation), or any forces could act insofar as they bear even a family resemblance to ordinary ones: in the end, such "forces" are no better than "phantom effects" and we are left with just another skeptical fantasy. At a minimum, defenders of

this example have not done the work needed to show that we are faced with a credible case of nonskeptical empirical equivalence. Were this one additional example to be accepted as genuine, however, it would not affect the status of my conclusion below.

8. Kukla (1993 5–6) accuses Laudan and Leplin of presuming that the case for underdetermination rests upon empirical equivalents alone. Leplin and Laudan (1993 16) deny the charge, but insist that their joining of the two doctrines was "not capricious," for "the philosophers whose derogations of the epistemic enterprise we have been concerned to redress (e.g., Quine and Rorty) . . . come to [the underdetermination thesis] . . . through [their belief in empirical equivalents]."

9. Sklar is, of course, in good company. Quine's classic (1975), for example, so often cited as providing *evidence* for an important underdetermination predicament, simply blusters: "Scientists invent hypotheses that talk of things beyond the reach of observation. The hypotheses are related to observation only by a kind of one-way implication; namely, the events we observe are what a belief in the hypotheses would have led us to expect. These observable consequences of the hypotheses do not, conversely, imply the hypotheses. Surely there are alternative hypothetical substructures that would surface in the same observable ways" (313).

10. Strictly speaking, of course, the case for recurrent, transient underdetermination requires only that there have routinely been (nonskeptical, nontrival) unconsidered alternatives *not effectively ruled out by* the evidence. I will try to show, however, that the stronger existence claim of unconsidered alternatives at least roughly equally well-confirmed by the available evidence is historically defensible, and it deflects any suggestion that such alternatives were ignored on evidential grounds rather than simply unconceived.

11. Notice that we cannot respond to these examples by noting that theories in the same general family or category as a later alternative (say "atomism") sometimes *had* already been entertained and/or dismissed by the time of an earlier theory's exclusive dominance, for our confidence in the truth of our present theories cannot survive an inductive rationale for thinking that present evidence likely supports a presently unconceived detailed version of a theory from an existing family or type just as well as it supports the present alternative we accept on the strength of that evidence. We will return to this point later in the more specific historical context of theories of generation and inheritance.

12. It is sufficient to ground the new induction, however, if we grant that the later theory is confirmed by the earlier phenomena *as described or conceived by the later theory itself* just as well as the earlier theory was confirmed by those same phenomena under its own description or conception of them.

13. Indeed, realists who attack the new induction by pointing out the significance of important shifts over historical time in the acceptance of auxiliary hypotheses, evidential standards, or theoretical conceptions of the phenomena themselves would seem to risk undermining the privileged position they ascribe to the corresponding auxiliary hypotheses, evidential standards, and theoretical conceptions of the present day.

2

Chasing Duhem

The Problem of Unconceived Alternatives

> I scored the next great triumph for science myself: to wit, how the milk gets into the cow. Both of us had marveled over that mystery a long time. We had followed the cows around for years—that is, in the daytime—but had never caught them drinking a fluid of that color. . . . deep in the woods I chose a small grassy spot and wattled it in, making a secure pen; then I enclosed a cow in it. I milked her dry, then left her there, a prisoner. There was nothing there to drink—she must get milk by her secret alchemy, or stay dry. . . . I stole away to my cow. My hand shook so with excitement and with dread failure that for some moments I could not get a grip on a teat; then I succeeded, and the milk came! Two gallons. Two gallons and nothing to make it out of. I knew at once the explanation: the milk was not taken in by the mouth, it was condensed from the atmosphere through the cow's hair. I ran and told Adam, and his happiness was as great as mine and his pride in me inexpressible.
>
> —from 'Eve's Autobiography' in "Papers of the Adam Family," Samuel Langhorne Clemens (Mark Twain), *Letters from the Earth*

2.1 Duhem's Worry: Eliminative Inferences and the Problem of Unconceived Alternatives

Worries about the possible existence of serious unconceived alternatives to our best scientific theories is certainly not new. Nearly a century ago the French physicist Pierre Duhem offered a characteristically lucid and provocative articulation of this challenge to the power of our scientific methods to uncover theoretical truths about the natural world:

> Between two contradictory theorems of geometry there is no room for a third judgment; if one is false, the other is necessarily true. Do two hypotheses in physics ever constitute such a strict dilemma? Shall we ever dare to assert that no other hypothesis is imaginable? Light may be a swarm of projectiles, or it may be a vibratory motion whose waves are propagated in a medium; is it forbidden to be anything else at all? ([1914] 1954 189–190)

What seems to have worried Duhem is the *eliminative* character of many important scientific inferences: often in science, perhaps even typically, we arrive at a decision to accept or believe a given theory because we take ourselves to have convincingly eliminated or discredited any and all of its proposed rivals or competing explanations of the available evidence. But as Duhem saw, such an eliminative inferential procedure will only guide us to the truth about nature if the truth is among these competitors in the first place.[1] And he wondered whether we had any good reason either to make or to reflectively endorse the assumption that it would be. This concern is given a special poignancy, of course, by he fact that we now regard light neither as a "swarm of projectiles" nor as a "vibratory motion whose waves are propagated in a medium."

The ensuing years have witnessed little progress in assessing the seriousness with which we should regard Duhem's worry. In part this is because it requires us to answer a question concerning which it is extremely difficult to acquire any compelling evidence. In fact, it requires us to answer the same question we raised in connection with recurrent, transient underdetermination in the last chapter: whether the group of hypotheses we haven't yet considered includes serious unconceived alternatives to contemporary accounts of nature offering equally (or at least reasonably) convincing explanations of the empirical evidence we have and therefore having an equally (or at least reasonably) claim to represent the theoretical truth about nature. And as we saw in the last chapter, the few philosophers of science who have considered this issue seem by and large to have simply assumed an answer to whether some version of this worry is serious or not and proceeded to beg the question against those who assumed otherwise.

In the previous chapter I suggested that the historical record of scientific inquiry itself offers us a compelling reason to think that Duhem's challenge is a serious one: a robust historical pattern of recurrent, transient underdetermination by previously unconceived alternatives would give us strong reason to believe that there probably are serious alternatives to even our best current theories that are presently unconceived, despite being well confirmed by the evidence available to us. In later chapters I hope to make the case for this claim in greater historical detail. Here I want to point out how the challenge itself, if it can be established in this way, should change the way we think about eliminative inferences and the scientific theories they are used to support.

In the context of Duhem's challenge, my general argument can be put as follows. Eliminative inferences are only reliable when we can be reasonably sure that we have considered all of the most likely, plausible, or reasonable alternatives before we proceed to eliminate all but one of them (or, in the limiting case, simply rest content with the lone contender). But the history of science shows that we have repeatedly failed to conceive of (and therefore consider) alternatives to our best theories that were both well confirmed by the evidence available at the time and sufficiently plausible as to be later accepted by actual scientific communities. Even more briefly, the historical record suggests that in science we are typically unable to exhaust the space of likely, plausible, or reasonable candidate theoretical explanations for a given set of phenomena before proceeding to eliminate all but a single contender, but this is just what would be required for such eliminative inferences to be reliable.

I wish neither to overstate nor understate what is at stake here. I am not suggesting that eliminative inferences are in general not to be trusted; indeed I very much doubt that we could get along without them. In many of the epistemic circumstances we encounter this kind of inference is a perfectly reliable tool, in particular when we are trying to choose among a fixed set of clearly delimited, exhaustive possibilities known in advance. To take a simple example, suppose I am playing bridge. For those who have never played the game, its relevant features are as follows. First, all the cards in a standard fifty-two-card deck are dealt out to each of the four players, so every card in the deck is in one of the four hands. One of these hands will become "the board," laid out face-up so that all players can see it and leaving only three players in the hand. Finally, players must follow suit: in a particular round of play, if a card of a given suit is played first, each other player must play a card of the same suit if she has one. Now suppose I can see that the queen of clubs is neither in my hand nor on the board. And in response to my club lead, one of the other players plays a diamond. I am now in a position to make a convincing eliminative inference to the conclusion that the queen of clubs *must* be in the hand of the other player. Or more cautiously, I might infer the disjunction of this claim and the more remote possibilities that the first player has played illegally, I have made an observational error, or the queen of clubs is missing from the game.

Of course, eliminative inferences do not always exclude alternatives by rendering them impossible: they can instead simply show that some possibilities are much more likely than others, as in the following example:

These tracks were made by a dog or a wolf.
Noone has ever seen a wolf this far south.
Therefore, these tracks were made by a dog.

In the bridge example, the first player played a diamond, rendering it impossible that she still (legally) holds the queen of clubs. By contrast, the fact that no one has ever seen a wolf this far south does not make it impossible that one hungry fellow has finally made the trip or that wolves in the area have eluded detection until now: it simply renders the possibility that the tracks were made by a wolf unlikely, leaving us with the comparatively much more likely (given the information we have) possibility that they were made by a dog. Here we exclude possibilities by showing that the evidence makes them implausible, rather than showing that the evidence simply rules them out altogether.

I will have little to say about how such inferences should be formally represented or understood. The bridge example seems most naturally regarded as a deductive enthymeme, whose premises include the claim that the queen of clubs must be in the hand of one or another player in the game. Of course, it could also be represented as an instance of abductive inference or *inference to the best explanation*, widely regarded as the central inferential tool of scientific inquiry, in which we infer that the best explanation of some further set of facts, data, or phenomena is true (or probably true, or approximately true, or probably approximately true, or some such): here the supposition that the second player holds the queen of clubs would be the best explanation of the known facts that it is missing from my hand and the board and that the first player responded to my club lead with a diamond. Conversely, the case of the dog/wolf tracks is a textbook example of an abductive argument; but it can easily be transformed into a deductive argument (cf. Musgrave 1988) by suitably supplementing and/or modifying its premises (i.e., adding "If no one has ever seen a wolf this far south, it is reasonable to believe that there are no wolves in this area," "If it is reasonable to believe that there are no wolves in the area, then it is reasonable to believe that the tracks were not made by a wolf," and so on) and weakening its conclusion ("Therefore, it is reasonable to believe that these tracks were made by a dog"). In general, it seems most natural to represent an argument as abductive or an inference to the best explanation when we recognize that the available evidence simply favors one of the stated possibilities without ruling the others out altogether, while a deductive construal seems more appropriate when the available evidence is strictly inconsistent with the various possibilities that it is used to exclude.

What is crucial, however, is that *either* representation of this sort of argument preserves its eliminative character: both reach conclusions by ruling out alternative possibilities until only one remains, in these cases from a plausibly exhaustive set of all the most likely alternatives. But I have suggested that this last feature is the source of a crucial disanalogy with theoretical science, for just this feature is lacking from the most distinctive scientific employment of eliminative inference. We are perfectly justified in regarding it as a remote possibility that the tracks were made by a heretofore undiscovered large mammal with a strikingly wolfish pawprint or that the queen of clubs has been quietly vaporized by anti-bridge forces beyond our ken. What I hope to show is that in scientific cases what evidence we do have should lead us to think that the chances that there are serious unconceived alternatives to our best theoretical explanations of natural phenomena are anything but remote.

For those who are inclined to think of a challenge to eliminative inferences in science as a challenge to the scientific use of abduction or inference to the best explanation, however, we might think of Duhem's worry in another way. In actual cases, any inference to the best (or only) explanation carries an implicit restriction: it is always an inference to the truth of the best (or only) explanation *we have managed to come up with so far*. Often enough, ignoring this implicit restriction is harmless because we are rightly confident in our ability to have exhausted the space of likely or plausible explanations in the first place (i.e., because these are the sorts of tracks that only dogs or wolves make) and this makes the restriction easy to ignore in formulating the structure of abductive arguments. But I hope to show that the restriction is of great significance indeed in the scientific case, because there we have abundant evidence that in past cases we have *failed* to canvas all of the likely, plausible, or well-confirmed theoretical explanations of the data before proceeding to eliminate alternatives. In the scientific case, I suggest, we have every reason to think that some of the very best theoretical explanations of the data are among those we have yet to even consider.[2]

It is important to be clear that the grounds for this claim do not depend upon our failure to consider radically skeptical scenarios, like the famous Cartesian "Evil Demon" described in the last chapter who devotes his energies to deceiving us about what the world is really like. As we noted there, such fantasies offer an equally powerful (or powerless) challenge to any knowledge claim of any sort, no matter how it is arrived at or supported. I explicitly grant that such radically skeptical possibilities are remote, or otiose, or of low prior probability, or whatever, and insist that we have grounds for concern about the reliability of eliminative inferences in the quite specific context of theoretical science nonetheless. The reason is that we have similarly specific positive evidence of our persistent inability to imaginatively exhaust the space of *scientifically serious* theoretical possibilities well confirmed by the evidence available at a given time: possibilities like Newtonian mechanics, Mendelian genetics, general relativity, and evolution by natural selection, and not like Cartesian Evil Demons. That is, I suggest that the historical record of scientific inquiry itself provides abundant evidence that the specific requirements for the reliable application of eliminative inference—the same requirements that really are satisfied (notwithstanding the possibility of Evil Demons) in many other applications of such inference—are routinely *un*satisfied in the context of theoretical science conducted by creatures who are cognitively constituted as we are. Thus, I offer no familiar sweeping Cartesian indictment of all knowledge whatsoever, nor even of all eliminative inference, but instead suggest that we are routinely using a perfectly legitimate inferential tool outside of the epistemic context in which it can be reasonably expected to uncover truths about the world.

As important as the point that not all eliminative inferences are suspect, is the point that not all scientific claims or beliefs are grounded in eliminative inference, for contemporary scientific inquiry is anything but methodologically monolithic. Instead it is an epistemically complex and heterogeneous enterprise in which various claims are advanced in various contexts on the strength of various kinds and degrees of evidential support, and I doubt that much interesting can be said that would apply to all and only "the claims of science" in general. Eliminative inference is certainly not the whole of what science does, nor is it reasonable to think that the warrant for every single scientific claim must be somehow infected by the reach of such eliminative foundations. For that matter, not even all *scientific* applications of eliminative inference are suspect. If we set out to test the hypotheses that alcohol consumption per capita among American high school students has increased, decreased, or remained steady over the past decade, I see no room for concern that we are failing to consider some important alternative possibility. Of course, there are methodological pitfalls involved in testing such hypotheses and we might well come to believe the wrong one, but this will not be because we have employed eliminative inference outside its domain of reliable application.

On the other hand, the set of scientific beliefs whose eliminative foundations seem most vulnerable to the challenge of unconceived alternatives will almost certainly include many or even all of those theories about remote or inaccessible domains of nature that form the very heart of our scientific conception of the world. Duhem was surely right to suggest that eliminative inferences play an especially important role in science, particularly in those parts of the sciences that seek to theorize about the fundamental constitutions of the various domains of the natural world and the dynamical principles at work in those domains: what we might call, for lack of a better term, our fundamental theories of nature. And it is when we theorize about such matters as the constitution of matter itself, the remote history of the Earth and its inhabitants, the most minute workings of our bodies, and the structure of the farthest reaches of the universe that we would seem to be in greatest danger of failing to conceive of serious alternative possibilities or even of what the space of such possibilities might look like, and I suggest that as a matter of historical fact this is just what we have repeatedly failed to do. I hope to show, then, that the evidence for the significance of the problem of unconceived alternatives is strongest just where the problem would matter most and just where it poses the most significant challenge to our scientific conception of the natural world: in our efforts to theorize about the most fundamental aspects of the constitution and dynamics of the various domains of nature.

It may be helpful to illustrate the crucial contrast between scientific beliefs that are and are not grounded on eliminative inferential foundations by means of some simple examples. The claim that a chunk of pure sodium will burst into flame when placed in water, for instance, is not known on eliminative grounds. Although one can imagine believing this claim only because it is implied by the best hypothetical explanation we have for some set of further phenomena, this is simply not our evidential situation. And the rather more direct evidence we do have in support of it gives us every reason to think that our account of the natural world will have to accommodate the experimental regularity we presently describe in this way even if the nomenclature and other descriptive apparatus of contemporary chemical theories is ultimately abandoned. Likewise in the case of what are sometimes called "phenomenological" or "experimental" laws of chemistry and physics, like the ideal gas law or the equations describing the coefficients of expansion for various materials. Such laws and equations may be redescribed or further refined, and their domains of application may be circumscribed, but the possibility of unconceived alternative explanations does not threaten to show that such claims of relationships between measured physical quantities are fundamentally mistaken, for these are not claims we believe simply because they offer (or are implied by) the best explanation we can think of for some further set of facts. In a similar fashion, the reasons we can give for believing that dinosaurs roamed the earth long ago and that tiny creatures invisible to the naked eye fill the world around us seem non-eliminative in character: fossilization is not a hypothetical postulated mechanism but a process we can study in action and we simply have no specific reason to doubt that its products in the remote past were any different from its present ones. Similarly, the ability of systems of combined lenses (like microscopes) to render visible those objects or features of objects whose minuteness makes them difficult or impossible to see in particular circumstances is easily and directly demonstrated (whether or not the theoretical hypotheses about the constitution of light and matter on which their construction depends are true) and again we simply have no reason to suppose that microscopes work differently or introduce sensory artifacts when their application is extended into domains where we could not see at all without them. Although skeptical alternative explanations of the evidence could be (and sometimes have been) *imagined* for any of these claims, it is not by ruling out the possibilities that fossils are a test of our faith from God or that microscopes mysteriously introduce amoeba-like perceptual artifacts beyond a certain level of resolution that we come to know the claims in question.

DISCARD
GARDNER HARVEY LIBRARY
Miami University-Middletown
Middletown, Ohio

By contrast, consider the claims that nothing can travel faster than the speed of light, that chemical bonds are constituted by the transfer or sharing of negatively charged electrons between positively charged atomic nuclei, that spiders and human beings share a common ancestor in the distant past, that the deformation of spacetime produced by massive bodies is responsible for their mutual gravitational attraction, and that self-replicating molecules emerged from a chemical soup to begin the history of life on Earth. Each of these claims is either itself a fundamental theoretical hypothesis or is a consequence of one. More importantly, the reasons we can offer for believing them would seem in each case to be limited to the fact that each of the fundamental hypotheses in question offers the most powerful and convincing systematic account we have for explaining, predicting and intervening with respect to a wide range of empirical phenomena (including in many cases phenomena unknown before the theory suggested their existence) and we can neither offer nor even imagine any alternative hypothesis whose performance in these respects would be equally impressive. It is this evidential situation to which the problem of unconceived alternatives is particularly germane, for I suggest that this is the position history reveals we have repeatedly occupied even when equally impressive alternatives making radically different claims about nature were in fact available, unconceived by us. If so, the fact that a scientific claim or belief is either part of or is implied by the best or only fundamental theory we have for explaining, predicting and intervening with respect to a particular scientific domain or set of empirical phenomena is simply insufficient to warrant the belief that it is therefore even approximately true.[3]

The heterogeneous character of the challenge thus posed by the problem of unconceived alternatives can perhaps also be helpfully illustrated by contrasting it with the most presently influential alternative to scientific realism: the *constructive empiricism* defended by Bas van Fraassen (1980). As we noted in the last chapter, van Fraassen is among those who are deeply worried by the prospect that even our best scientific theories might have empirical equivalents sharing just the same observational implications. As a consequence, he urges us to believe only the claims our theories make about *observable* phenomena: just the claims they hold in common, of course, with any empirical equivalents they might have. It is widely argued that this choice of dividing line is indefensible, not only because the distinction between observables and unobservables is itself vague, specious, or nonexistent, but also because it is epistemically unmotivated, insofar as van Fraassen offers no reason for doubting the truth of what theories say about unobservables that could not equally be invoked to doubt the truth of what they say about observable but unobserved phenomena. I suspect that this complaint misses the point of van Fraassen's voluntarist epistemology: his position seems to me to be that a reflective endorsement of the successful practices of scientific inquiry requires no more of us than the belief that what our theories say about observables is true, but also requires no less; that is, that belief in the empirical adequacy of our theories is the *minimal* degree of epistemic commitment required to endorse the practices of science itself and is therefore the level that prudence recommends to thoughtful defenders of scientific inquiry.

Nevertheless, I do not share van Fraassen's generalized fear of commitment. My interest lies in finding out how *much* we can actually know, not how little we can get away with believing while using or doing science; indeed, seen in this light, van Fraassen's constructive empiricism seems, in words Karl Popper once used to deride instrumentalist alternatives to realism generally, "a narrow and defensive creed" (1963 103). If my earlier claims are right, observability is important, but only derivatively, for it seems natural enough to think not only that it is as we are pushed farther and farther from the evidential resources of immediate observation and experience that we are increasingly forced to justify beliefs by eliminative competition among hypothetical possibilities, but also that our grip on the contours of the space of such alternative possibilities becomes progressively less secure, and thus the danger of our failing to conceive of serious alternative theoretical explanations for the phenomena when they exist becomes correspondingly more acute. But we neither invariably reason eliminatively about unobservables, nor invariably reason otherwise about observables, and it is the application of eliminative inference outside its domain of reliable operation, not observability as such, which represents a legitimate source of concern or a reasonable ground for withholding our belief from the claims of theoretical science. The difference is perhaps well illustrated by the existence of microscopic organisms, a phenomenon for which I have suggested we have convincing evidence of a non-eliminative character, but about which van Fraassen's commitment to the epistemic significance of observability forces him to remain agnostic instead. Conversely, our characterizations of and beliefs about many observable entities (e.g., chemical elements, evolutionary adaptations, supernovae) would seem to be routinely grounded in and bound up with just those sorts of fundamental theoretical conceptions of the natural world that stand most forcefully challenged by the problem of unconceived alternatives. It does seem in general that eliminative inference is both especially important and especially suspect in trying to uncover truths about domains of nature remote from observation, where our access to the kinds of evidence needed for other sorts of inferences or even to constrain the space of theoretical possibilities is so much more limited. But for all that there is nothing especially suspicious about scientific claims regarding unobservables *per se*. Indeed, the historical challenge is correctly focused not on our beliefs about entities of a particular sort (i.e., "unobservables"), but instead on beliefs *arrived at* or *justified* in a particular way. More specifically, there is something suspicious about *any* claim arrived at eliminatively in that particular set of epistemic circumstances (it so happens) in which theoretical science is routinely forced to operate.

This should also help to clarify why the problem that concerns us here neither constitutes nor collapses into a disguised version of global skepticism. For one thing, the grounds for my beliefs that I am now wearing pants and that I had eggs for breakfast this morning are not eliminative (or not obviously eliminative anyway). But even if they were (and even for those who think that we cannot justify such beliefs unless we eliminate perverse Cartesian possibilities) the reasons I will offer for skepticism about the reliability of eliminative inference in theoretical science simply are not grounds for skepticism about the employment of eliminative inferences in general, much less for skepticism about all beliefs whatsoever. I will point to a specific history of our repeated failures to exhaust the space of serious scientific alternative possibilities, and there is simply no comparable history available of failures to conceive of and therefore consider presumptively plausible alternative explanations for the evidence supporting beliefs like that I am now wearing pants or that I had eggs for breakfast. The difference is anything but mysterious, of course, insofar as the former sorts of alternatives are neither easy to conceive (particularly in the requisite detail) when they exist, nor is their plausibility at all easy to assess even when they can be conceived. But in any case, the fact that any ampliative form of inference *might* go wrong is neither news nor grounds for any startling conclusion; what is significant is the fact that we seem to be making routine use of a particular form of such inference outside of the epistemic context in which it can be expected to operate reliably, or rather, in conditions in which we can know for a fact that it will be unreliable.

A similar point applies to many other knowledge claims that are arguably theoretical in character. For example, in a justly famous discussion W. V. O. Quine (1955) argues that desks and other objects of everyday experience are no less posits accepted because they help explain our sense-data than are molecules. But I have suggested that an important difference remains. While there are certainly possible alternatives to the commonsense ontology of an external world full of physical objects (e.g., phenomenalism), the historical record does not find us continually discovering previously unconceived alternatives of this sort that are ultimately plausible enough to attract entire communities of sincere proponents, while I suggest that we do find just this historical situation in the case of theoretical science. Thus, our scientific theories share a demonstrated historical vulnerability to the problem of unconceived alternatives that Quine's hypothesis of "the bodies of common sense" simply doesn't share.[4]

Finally, the specificity of this charge points to a further important difference between my concerns and some more traditionally philosophical efforts to challenge the veridicality of scientific claims: I am not reaching beyond or outside of science itself for evidence of some supposedly higher or purer kind with which to sit in global judgment on the scientific enterprise as a whole. With those philosophical naturalists who emphasize the essential continuity of philosophical and scientific efforts to acquire knowledge, I hold that there is only good and bad evidence, not higher and lower evidence or scientific evidence and some other kind. Indeed, I expect my argument to be congenial to at least naturalists of this sort, if not the more dogmatic variety who insist on taking the deliverances of contemporary scientific theories for granted as providing not only the starting point but also the boundary conditions for serious philosophical inquiry. What I suggest is that we are already in possession of abundant evidence of a perfectly ordinary empirical sort concerning our repeated failure to exhaust the space of serious and well-confirmed theoretical alternatives available at any given time. This repeated failure constitutes an evidential *constraint* that will have to be satisfied by any convincing account of ourselves (naturalistic or otherwise) as cognitive agents, but the constraint alone will suffice to rationally preclude us from believing the deliverances of our own eliminative inferences in the case of fundamental theoretical science to be literally or even approximately true: this is a case of the naturalist learning something from the world about the reliable scope of one of her own learning processes.[5]

Thus, a sober consideration of the problem of unconceived alternatives casts suspicion neither on all eliminative inferences, nor on all scientific beliefs, nor even on all scientific applications of eliminative reasoning. Instead it encourages us to distinguish claims or beliefs according to the *kinds of evidence* we have for them, and it counsels skepticism about all and only claims arrived at or justified eliminatively *when we have good reason to doubt that we can exhaust the space of plausible alternative possibilities*. It will therefore be quite difficult to anticipate the verdict that the problem recommends concerning particular cases of scientific belief in advance of detailed investigation: the limited skepticism thus motivated should certainly not extend to every scientific claim or hypothesis and may even have different force as applied to the scientific exploration of different domains.[6] Nonetheless, the systematic evidence available from the historical record seems to promise to give us good reason to doubt our ability to exhaust the space of plausible alternative possibilities in the context of fundamental theoretical science quite generally, thus posing a distinctive general challenge to virtually all of those fundamental theories concerning remote domains of nature that lie at the heart of the contemporary scientific conception of the natural world. And of course such a picture of the scientific enterprise, on which the fundamental accounts of nature offered by even its most successful theories cannot be regarded as even approximately true, stands at a considerable distance from much scientific and philosophical orthodoxy.

2.2 Confirmation: Holism, Eliminative Induction, and Bayesianism

I have suggested that the problem of unconceived alternatives does not pose a serious challenge to every eliminative inference, nor even to every scientific application of eliminative reasoning, and certainly not to every scientific belief: in many cases of eliminative inference we can be confident that our grasp of the plausible alternatives is indeed exhaustive, and there are many scientific beliefs whose supporting evidence simply does not seem to be eliminative in character at all. But this latter claim required us to distinguish beliefs confirmed by their systematic predictive and explanatory power, such as the claim that the deformation of spacetime is responsible for gravitational attraction or the claim that spiders and humans share a common ancestor, from beliefs confirmed in some more "direct" or less mediated way, such as the ideal gas law or the claim that pure sodium will burst into flame when placed in water. And any such distinction will seem profoundly misguided to those philosophers of science who have embraced an influential view of the general relationship between our beliefs and the evidence we take to support them entitled *confirmational holism*.

The holist position begins from the insight that our beliefs (including especially our scientific theories) do not carry empirical implications all by themselves, but do so only in conjunction with others: the belief that the universe began with a "big bang," for instance, only carries implications about what I should expect to find here and now in conjunction with further beliefs about what the present effects of this big bang should be (e.g., cosmic background radiation) and about how these effects produce or influence observable phenomena (e.g., the background "snow" on a television set that is receiving no broadcast signal). If a theory's predictions turn out to be mistaken, the holist points out, this forces us to give up *something*, but this might well be our beliefs about the present consequences of the big bang or about how cosmic background radiation interacts with television sets rather than our beliefs about the origin of the universe. From this the holist concludes that no beliefs are straightforwardly verified or refuted by particular experiences or collections of experiences and that all beliefs are confirmed only by serving as part of a collection that accounts well for our perceptual experience on the whole. In Quine's famous image, the interconnected web of any person's beliefs makes contact with her perceptual experience only at its periphery, and this supposedly ensures in turn that all of the claims involving a given belief (including a given scientific theory), and indeed all of human knowledge itself, confronts the tribunal of experience only as a corporate body. On this holist view it would seem that all scientific beliefs, indeed all beliefs whatsoever, are tested by the evidence in an identical fashion, and thus that there is no room for a distinction between those confirmed by their systematic predictive and explanatory power and those confirmed in some other way.

There is little doubt, I think, that the holist arguments from the interconnections among our beliefs to the in principle possibility of preserving any belief by giving up others illustrate something important about the nature of scientific testing and confirmation. Nonetheless, the extreme holist claim that all of our beliefs are confirmed solely by their inclusion in an interconnected web that accommodates experience well on the whole may already have seen its day. More recent epistemology and philosophy of science has witnessed calls for (and some proposals of) more nuanced accounts of confirmation recognizing the differential character of the evidence in support of different kinds of scientific claims and a tighter relationship than the holist allows between particular claims and particular bits of empirical evidence. Indeed, in later chapters we will see how some scientific *realists* have sought to use the implausibility of such extreme holism to try to defend their commitment to the claims of current theories in the face of historical challenges. I will have little to contribute here to the substantive development of such alternatives to confirmational holism or to any general account of how particular pieces of evidence bear more directly on one part of a theory than another. But I nonetheless suggest that the examples we have considered of scientific beliefs that are and are not grounded on eliminative inferences themselves suffice to illustrate why any account ignoring the substantial differences between these cases will be missing something significant about the heterogeneous character of scientific confirmation.

Of course we might ultimately come to agree with the holist that the sorts of confirmational differences noted above are only a matter of degree, but even if so this difference of degree will itself be one of great evidential significance. In fact, the difference in question might seem to correspond quite naturally to the extent or the respects in which the available evidence constrains the space of possible alternative explanations facing us: our microscopic observations and fossilized skeletons seem to rule out the serious possibility that amoebae don't (or that dinosaurs didn't) exist at all in a way that the serious possibility of alternative explanatory accounts of the minute structure of matter, of the relationship between dissimilar species of organisms, or of the source of gravitational attraction simply have not been (and perhaps cannot be) similarly ruled out, *even when we can't say anything about what the latter alternatives might be like.* Characterizing the evidential difference between such cases strikes me as one of the hardest problems facing the contemporary philosophy of science, but if I am right to suggest that the problem of unconceived alternatives poses the most serious challenge to believing the claims of contemporary scientific theories, sorting out this difference will prove to be important work worth doing. In the meantime, of course, for those who continue to embrace confirmational holism the problem of unconceived alternatives will be even more important and more worth our sustained attention than I have suggested thus far. For if the extreme holist turns out to be right after all, this simply means that the reach of the problem of unconceived alternatives is considerably *longer* than it appears at first glance and that many *more* of our scientific beliefs are at risk from the possibility of unconceived alternatives than our earlier survey of putatively divergent examples would suggest. Of course, it will still be only our theoretical scientific beliefs that are systematically at risk in this way, however, as it is only these for which we have found evidence of a general historical vulnerability to a significant version of the problem of unconceived alternatives. As we noted above, for instance, there is simply no comparable history of serious unconceived alternatives to the hypothesis of "the bodies of common sense."[7]

In any case, even if we resist the radical confirmational holist's suggestion that all scientific confirmation will remain vulnerable to the problem of unconceived alternatives, it is certainly worth noting the increasingly widespread recognition of the prevalence and importance of explicitly eliminative inferences in the context of science itself. Recent years have witnessed sophisticated and systematic treatments by philosophers of science of the central role played by eliminative inferences both in establishing particular scientific claims and as a general inferential strategy for scientific inquiry (see Earman 1992 chap. 7; Kitcher 1993 chap. 7; Norton 1993, 1995, 2000; Achinstein 2002). Despite the sophistication of these discussions, however, they have typically offered little in the way of either a response to the problem of unconceived alternatives or a reason to think we can afford to dismiss it. Discussing Jean Perrin's efforts to experimentally confirm the existence of atoms, for instance, Peter Achinstein recognizes that the arguments Perrin relied on to eliminate alternative possible causes of Brownian motion are vulnerable to the possibility that he has not considered all the serious candidate causes, but has nothing substantive to offer in reply, suggesting simply that "the burden of proof is on the critic" and that "[s]ince Perrin cited and eliminated various possible causes, it is, I think, up to the critic to say what other possible causes he should have eliminated, given his information" (2002 479). This is simply to miss or ignore the possibility that we might be in a position to reasonably doubt Perrin's (or our own) ability to exhaust the space of plausible alternatives without being able to specify a particular alternative he has failed to consider.

By way of contrast, John Norton (1993, 1995, 2000) and John Earman (1992 chap. 7) have offered impressively detailed accounts of some important cases from theoretical physics in which scientists' uses of explicitly eliminative inferences have involved efforts to restrict and regiment the space of theoretical alternatives under consideration in such a way that the empirical evidence really can systematically eliminate all but a single contender. Of course, their cases are drawn exclusively from a single scientific domain and give us little reason to suppose that we will in general be able to exhaustively characterize even well-defined parts of the space of theoretical possibilities in this way in most scientific fields. But far more problematic is the fact that these efforts have invariably made use of substantive assumptions about the world (or equivalently, about the form and/or content of the correct theory for a given domain of nature) in order to remove indefinitely characterized and/or infinitely large parts of the space of possibilities from consideration and to restrict our eliminative attention to a relatively small and well-behaved part of the space of remaining theoretical alternatives. While the substantive assumptions used for this purpose are certainly ones that seemed natural and perhaps even unavoidable to the scientists who appealed to them, we will see in later chapters that similar assumptions about the world or about the form and content of the correct theory have routinely seemed natural, reasonable, and/or unavoidable to practicing scientists only *because* they failed to consider the serious alternative possibilities that would ultimately be embraced by later theorists. That is, while there are certainly cases of eliminative inferences in which we can justify restricting our attention to some small part of the space of possibilities (e.g., "these tracks were made by a dog or wolf"), our historical investigation will suggest that in the case of fundamental theoretical science it is often a *consequence of* our failure to conceive of the serious alternative possibilities that do in fact exist that we embrace the substantive assumptions needed to restrict the space of theoretical alternatives under consideration to a comparatively small and/or well-behaved set. Thus, showing that eliminative inference can be effectively defended in theoretical science after we use substantive empirical assumptions to radically restrict the space of alternative theoretical possibilities we are considering is simply to presuppose and not to provide a solution to the problem posed by the possibility of serious unconceived alternatives.

The leading general approach to theoretical confirmation in the contemporary philosophy of science, *Bayesian confirmation theory*, occupies a similar position. Bayesians have long been sensitive to the worry that the truth of the matter might not be among the hypotheses under explicit consideration at a given time in the course of scientific inquiry. Abner Shimony (1970) famously responded to this worry by introducing the memorable device of a "catch-all" hypothesis into the Bayesian confirmational machinery: that is, an hypothesis representing the disjunction of all other possibilities *not* explicitly under consideration at a given time, whether this is because they are unconceived, remote, or ridiculous. But to protect the Bayesian account from the threat of skepticism, Shimony proceeds to give preferential treatment to hypotheses that have been "seriously proposed" by some practicing scientist and are therefore not in the catch-all.[8] This, of course, is to prejudge the seriousness of the problem of unconceived alternatives by simply assuming that merely unconceived or unsuggested hypotheses are no more likely to be true than frivolous Cartesian skeptical possibilities, and Shimony acknowledges that he can justify treating merely unsuggested alternatives on an equal footing with skeptical fantasies in this way only as a "counsel of desperation" (1970 81). That is, he argues that "unless we act *as if* good approximations to the truth will occur among the hypotheses which will be seriously proposed within a reasonable interval, we are in effect despairing of attaining the objective of inquiry" (1970 132 original emphasis). Of course the problem of unconceived alternatives concerns whether we have some good reason to *actually believe* that the truth of the matter or some reasonable approximation to it has appeared within the set of seriously proposed hypotheses; in the absence of any such reason, proceeding on the assumption that it has is simply wishful thinking. Furthermore, refusing to assume that the truth must be among the theoretical alternatives actually proposed at or by a given time despairs of "attaining the objective of inquiry" only if we have already decided that scientific realism is the only view of the matter we are prepared to accept.[9]

More recently, Wesley Salmon (1990) has pointed out that Bayesians can effectively evaluate the *comparative* confirmation of two hypotheses without presuming anything about what the catch-all hypothesis is like or how likely the truth of the matter is to be found there. Salmon acknowledges the "utter intractability" of evaluating the likelihood conferred on the evidence by the catch-all (that is, how likely it is that we would have the available evidence we do if the catch-all and our background knowledge are true), but he points out that this troublesome quantity can be made to drop out of the equations Bayesians use to evaluate the relative confirmation conferred by the evidence on one actually proposed hypothesis as compared to another. As he also clearly recognizes, however, this is simply beside the point in any dispute concerning scientific realism, for at issue there is whether or not we should believe that the best-confirmed theory emerging from such an eliminative comparison *actually represents the truth about nature or not.* And here Salmon acknowledges that Bayesians are simply stuck: the resources at the Bayesian's disposal "provide no evaluations of individual theories; they furnish only comparative evaluations" (1990 281). There is simply no way to assign an absolute probability or level of confirmation to the theory without solving the problem of estimating the likelihood conferred on the evidence by a catch-all hypothesis of unknown content and constitution. Perhaps equally important is the fact that we cannot obtain the Bayesian's reassuring results showing that initially divergent assignments of prior probabilities to hypotheses will eventually converge (the "washing out" or "swamping" of the priors by the evidence) without a responsible estimate of this likelihood (Salmon 1990 270, see also

Earman 1992 168–169). In a useful discussion, John Earman concedes Salmon's characterization of the evidential situation, and he concludes that Bayesians will simply *have* to find a way to characterize spaces of serious theoretical alternatives exhaustively if they are to deliver the sort of absolute, rather than merely relative, judgments of confirmation required to responsibly decide the likely truth or falsity of our scientific theories (1992 171–172).

Thus, neither the philosophers of science most sensitive to the crucial role played by eliminative inferences in science, nor those especially interested in theoretical confirmation itself have resolved or even directly engaged the issue that I have suggested matters most for the controversy over scientific realism: whether we ought to believe that there are scientifically serious alternatives to our fundamental theories of nature among the present space of possibilities we have not yet conceived of or considered. That is, they have been simply unable to effectively address the question of whether our consideration of the space of alternative theories in scientific contexts is in general sufficiently robust to allow the eliminative inferences we rely on to serve us as an effective means of reaching the truth. Thus, their careful examples of the use of eliminative inferences in fundamental scientific theorizing and their clear recognition that the Bayesian's formal tools will not help resolve the issue simply highlight the urgency of the problem of unconceived alternatives and the central importance of deciding whether there are indeed serious theoretical alternatives presently unconceived by our own scientific communities or not.

2.3 Pessimism Revisisted

Before turning to the sort of historical evidence I have suggested bears on this question, I also want to contrast the problem of unconceived alternatives with the pessimistic induction's more famous use of historical evidence to challenge scientific realism. In particular, I want to point out why the most important grounds that have been offered for resisting or criticizing the classical pessimistic induction simply do not apply to the challenge posed by the problem of unconceived alternatives itself. As we noted in the last chapter, the appealing simplicity of the pessimistic induction's "fool me once, shame on you; fool me twice, shame on me" argumentative strategy leaves it vulnerable to the obvious retort "that was then and this is now." Thus, scientific realists are quick to point out differences in the breadth, precision, novelty, or other important features of the predictive and explanatory accomplishments of past and present theories and to claim that these differences invalidate the pessimistic induction's attempt to project from past to present cases. And even if we cannot say why just *these* features ensure the truth of the theories that enjoy them when others that equally excited our initial admiration and credence turned out not to do so, we know that such differences sometimes do make a difference to the legitimacy of a proposed inductive projection. So there is at least some justice in the realist's suggestion that these differences should protect our own scientific theories from invidious comparison with their predecessors.

For this reason, it is important to point out that this reply to the pessimistic induction simply does not apply to the problem of unconceived alternatives or to the new induction that supports it. This is because the latter arguments concern the *theorists* rather than the *theories* of past and present science. That is, they point out not that past theories have ultimately been found to be false or otherwise wanting in some way despite sharing many virtues with the best theories of our own day, but instead that they were at one time *the best or only theories we could come up with*, notwithstanding the *availability* of equally well-confirmed and scientifically serious alternatives. Thus, the problem of unconceived alternatives and the new induction suggest not that present theories are no more likely to be true than past theories have turned out to be, but instead that present theorists are no better able to exhaust the space of serious, well-confirmed possible theoretical explanations of the phenomena than past theorists have turned out to be. And neither the force of this concern nor the validity of projecting it into the future is in any way mitigated by pointing out that many present theories differ in important and systematic respects from those of the past.[10]

Of course it remains possible to try to challenge the legitimacy of projecting the new induction from past to present cases in a similar fashion, but to do so would require some reason to believe that present scientists or scientific communities have somehow managed to acquire the ability to exhaust the space of serious and well-confirmed theoretical explanations for a given set of phenomena. That is, to pursue the analogous line of criticism against the new induction, its opponents would have to argue that some important feature of the institutions or practice of current science renders its theorists dramatically more proficient than their predecessors at exhausting the space of serious, well-confirmed possible theoretical explanations for a given set of phenomena. This suggestion is certainly not absurd, but defending it will require a substantive further argument that promises to be difficult to make: it is far from obvious that any such feature systematically distinguishes today's scientists or scientific communities from those of the past. In any case, much more will be required than simply pointing out the sorts of significant differences between some present and past theories to which realists have traditionally appealed to try to insulate current science from historical challenge. The unparalleled breadth, predictive power, precision or other substantive epistemic virtues of some present theories go no distance whatsoever toward showing that we have somehow acquired the ability to exhaust the space of scientifically serious alternatives: indeed, it is in part because these very theories, with the genuine advantages over their predecessors they really do enjoy, were themselves *previously* part of the space of unconceived alternatives that we seem to have every reason to believe that serious, well-confirmed alternatives to our own theories remain presently unconceived.

Of course, we need not see the problem of unconceived alternatives as supported solely by a simple enumerative induction projecting directly from past cases of failure to conceive of serious theoretical alternatives to future ones in any case. We might instead think of ourselves as marshalling evidence from the historical record for a quite general claim about human beings as cognitive agents: that we are not good at conceptually exhausting spaces of serious alternative possibilities with the sort of amorphous and indefinite contours characteristic of those from which we draw our fundamental theories of nature. While it seems possible to imagine cognitive supercreatures who are adept at conceiving of all possible theoretical explanations for a given set of phenomena (or at least all those that they and/or their successors might regard as scientifically serious), the evidence suggests that we are simply not cognitive agents of this kind. This suggestion goes further than either the pessimistic induction or the new induction alone, of course, in offering a natural explanation for the patterns we find in the historical record of scientific inquiry: it suggests that we repeatedly find our successful scientific theories replaced by others and that we repeatedly fail to conceive of well-confirmed theoretical alternatives that will be later embraced *because* we are creatures whose cognitive constitutions are not well suited to the task of exhausting the kinds of spaces of serious candidate theoretical explanations from which our scientific theories are drawn.

In this way, I view the problem of unconceived alternatives not as competing with the traditional challenges of underdetermination and the pessimistic induction so much as bringing out what was most significant and compelling about those challenges to begin with. The new induction provides a pure and simple inductive argument for the claim that we have repeatedly and characteristically occupied a significant underdetermination predicament, failing even to conceive of theoretical alternatives well confirmed by the available evidence that would later be embraced by actual scientists and scientific communities. The problem of unconceived alternatives goes still further, suggesting a natural explanation for the historical patterns recorded in both the new induction and the pessimistic induction: the realization that our cognitive constitutions or faculties are not well suited to exhausting the kinds of spaces of serious alternative theoretical possibilities from which our fundamental theories of nature are drawn. If this claim can indeed be supported by a detailed investigation of the historical record, it not only offers substantial reason to believe that we cannot trust eliminative inferences in the context of fundamental scientific theorizing, but also offers the most convincing affirmative reason we have to expect the patterns recorded in both the classical pessimistic induction and the new induction over the history of science to persist into the future.

Insofar as the problem of unconceived alternatives offers an explanation for the historical patterns recorded in the pessimistic induction and new induction, however, it is natural to wonder whether the explanation it suggests is not itself vulnerable to the very challenge it poses. That is, is this not itself an eliminative inference to one (the best?) among a number of possible explanations of the *historical* evidence, and, if so, shouldn't we be concerned about possible alternative explanations of that same historical evidence that have not yet been conceived or considered?

Here we would do well, however, to remember the earlier lesson that not all eliminative inferences are equally vulnerable to the problem of unconceived alternatives itself: the evidence we have does not support a blanket challenge to all eliminative inferences or to eliminative inferences in every epistemic context. I have suggested that the historical record offers us a succession of cases in which alternatives to our scientific theories remained unconceived by thinkers who would have or should have taken them seriously had they but had the opportunity to consider them. This is simply not the situation we face in the case of the relevant historical evidence. We do not encounter a long record of failures to conceive of serious alternative accounts or explanations of the historical evidence, so we have no similarly specific reason to distrust the best explanation we can offer. It is not that the problem of unconceived alternatives is a special problem, by fiat, only about theoretical science, but rather that it is only a serious concern where the conditions needed for the successful application of eliminative inference can be shown to fail, and we have no specific reason to regard the problem's own explanation of the historical evidence as a context of this sort. Thus, even if we understand the explanation offered by the problem of unconceived alternatives as itself an instance of eliminative and/or abductive reasoning, the eliminative or abductive inference in question will not be one that the problem itself offers any clear reasons to distrust.[11]

Of course, whether or not the explanation of the historical evidence offered by the problem of unconceived alternatives is accepted, we need not rely on this explanation to establish the seriousness of the challenge posed to scientific realism by the problem itself. The historical pattern recorded in the new induction alone suffices to establish that conclusion by a more simple and direct inductive route. *Whatever* the reason or explanation, I suggest that past scientists have in fact repeatedly and characteristically failed to conceive of the serious alternative possibilities that would ultimately be embraced by their successors. Moreover, we find that our failure to conceive of a given theoretical explanation immediately or within a given time frame offers no good reason to think that it will ultimately prove unfruitful, unsuccessful, or otherwise unappealing to our future selves or scientific communities. And even this simple challenge to scientific realism remains substantially harder to resist than that offered by the classical pessimistic induction, as it is much easier to point to characteristics of present theories that might distinguish their expected fortunes from those of past theories than it is to identify or even suggest any characteristics of present theorists that should lead us to think they are significantly better able to exhaust the space of theoretical possibilities than were the greatest scientific minds of the past. Thus, even if we think that there are indeed important epistemic differences between past theories and many of those of the present day, the new induction alone can entitle us to the conclusion that there are nonetheless probably serious, well-confirmed alternatives to even these more impressive latter theories that remain unconceived by us.

This discussion makes clear, however, how much ultimately depends on my claim that the historical record supports the new induction and the problem of unconceived alternatives in the way that I suggest. It is therefore time (perhaps well past time) to turn to a more direct consideration of this evidence. Since my broadest claims concern the entire history of scientific inquiry, it would be hopeless to try to fully defend them here, even if this were not well beyond my competence. What I can and will do, however, is take up at least one significant set of theoretical developments in the history of science and try to use it to illustrate the broader pattern I claim is characteristic of the history of scientific inquiry more generally. Over the course of the next three chapters I will take up a particular period of the history of our scientific theorizing in a particular domain of nature and try to show that that it exhibits the pattern I suggest: we will find that past theorists repeatedly failed even to conceive of avenues of theoretical explanation for the evidence available to them that were sufficiently scientifically serious as to be actually embraced by their successors. And unless we find some reason to think that this pattern depends on idiosyncrasies of the personalities or period involved in this particular case of fundamental scientific theorizing, even this single series of historical episodes may go a considerable distance toward showing that we are in possession of a quite general challenge to scientific realism about our fundamental theories of nature.

Notes

1. Note that this worry depends *in no way* on Duhem's much more famous commitment to confirmational holism (discussed below) and consequent belief that as a matter of principle no amount of experimental evidence suffices to rule out a given theory.

2. Note that this is quite different from the more familiar objection that in inference to the best explanation "our selection may well be the best of a bad lot" (Van Fraassen 1989 143): even if our theory is the best of a very good lot, the

worry is that there may be other, equally good alternatives which remain presently unconceived.

3. The suggested distinction between beliefs confirmed by their evident predictive and explanatory powers and beliefs confirmed in some more direct or less mediated way will seem suspect to those philosophers of science who are sympathetic to confirmational holism. We will consider this holist reservation briefly in the next section.

4. In the final chapter we will return to this point and its important implications for the coherence of any alternative to scientific realism.

5. Such an argument would be unconvincing if it depended on a prior theoretical account of our cognitive functioning that was itself eliminatively established, but that is clearly not the case. It would nonetheless be *self-defeating* if its claims about our cognitive limitations and their significance were themselves arrived at eliminatively in epistemic circumstances like those that bring the reliability of eliminative inference in theoretical science into question. Below I consider (and reject) the suggestion that the argument reaches its conclusions about the limits of our own eliminative inferential abilities in a way that renders it vulnerable to the very challenge it articulates.

6. I suspect that this is an important part of what accounts for the persistent appeal of the suggestion that questions about the justifiability of scientific beliefs or commitments can only be productively pursued on a piecemeal or case-by-case basis, although I am, of course, trying to provide a systematic rationale and principled methodological foundation for at least one aspect of such piecemeal approaches.

7. In fact, not only would the problem of unconceived alternatives not be undermined by an embrace of such radical confirmational holism, but our efforts to distinguish the problem from global or Cartesian skepticism offer a substantive lesson for the distinctively holist claim of underdetermination that has been widely influential in philosophy (e.g. Quinean epistemology) and in the study of science more generally (e.g. the Sociology of Scientific Knowledge (SSK) and feminist critiques of science) according to which our response to experimental *falsification* or *disconfirmation* is systematically underdetermined by the evidence. Such holists have famously insisted that when we are faced with a theory's failed predictions or mistaken empirical implications there are always innumerable ways to revise the theory and/or the web of background beliefs on which its implications also depend in order to render them consistent with our new evidence. This holist claim of underdetermination raises a fundamentally distinct epistemological issue, as it concerns an alleged inability to decisively *disconfirm* known theories rather than confirm them against unknown alternatives, but it depends on a similar claim about the universal availability of alternatives: to wit, that there are always innumerable ways available to revise an hypothesis or our background beliefs so as to render that hypothesis consistent with or confirmed by the evidence we have. And holists have sought to defend this presumption by appeal to strategies with what should now strike us as a familiar air of Cartesian desperation. Quine famously suggests, for instance, that we are always free to preserve a favored theory by pleading hallucination in light of any evidence whatsoever, and although he repeatedly insisted that his intention was to make a purely logical point without epistemological significance,

defenders of holist underdetermination have not always followed him in this assessment. Sociologists of science have instead offered a somewhat shrill insistence on the blanket claim that the legitimacy of absolutely any modification of an hypothesis or any challenge whatsoever to our background beliefs (about the evidence, other theories, principles of scientific methodology, legitimate sources of knowledge about the natural world, or whatever) is constrained solely by ongoing "social negotiation" between the parties involved and that the outcome of such challenges is therefore determined largely if not exclusively by the comparative social resources (e.g. power or authority) of the groups who find it in their interests to advance them or the theories they protect. If there is a genuinely consequential version of this holist challenge, I suspect that it would be more effectively defended by an analogue of the new induction asserting the historical ubiquity of *manifestly plausible* modifications of theories and challenges to background assumptions *that were themselves typically unconceived before the anomalous evidence arrived* than by appeal to these maneuvers, which threaten to bury any significant case for the distinctively holist form of underdetermination in a point of no more than Cartesian epistemological significance. (We will see at least one genuine instance of this sort of previously unconceived modification taken in response to threatened falsification in chapter 4, when we consider Darwin's response to his cousin Francis Galton's experiments with blood transfusions in purebred rabbits.) Of course this leaves entirely open the question of the extent to which existing case studies in the Sociology of Scientific Knowledge and/or feminist science studies have succeeded in showing what they claim to have shown about the crucial role played by political and social factors in determining the course of consensus and further inquiry in many particular scientific episodes.

8. That is, Shimony's "tempered personalism" directs us to assign to the individual disjuncts of the catch-all prior probabilities "which are generally many orders of magnitude smaller than those" assigned to actually proposed hypotheses (1970 131).

9. Shimony does go on to suggest that if we help ourselves to the *results* of inquiry grounded on the optimistic methodological presumption that the truth has appeared among the seriously proposed alternatives, the presumption itself can then be supported in turn by the resulting facts about ourselves as cognitive agents and the history of our scientific inquiry. Although he suggests that the evident circularity here is not vicious, Shimony's suggestion nonetheless depends on his own convergentist reading of the history of science, which notes not only the striking successes of many individual hypotheses but also claims a record of progressive theory replacement in which "the continuity of scientific knowledge is to some extent maintained by the existence of 'correspondence' relations between old and new theories" (1970 143–144). This claim and the use Shimony makes of it are vulnerable to the argument I will make against the probative significance of similar realist readings of the historical record in chapters 6 and 7.

10. That is, pointing out such characteristics of present theories does not provide any reason to challenge the legitimacy of the new induction's inductive projection from past to present cases, or the consequent claim that there are probably serious and well-confirmed alternatives to our own theories that are presently unconceived. This is of course quite different from suggesting that

some characteristics of some present theories give us grounds for confidence in their truth *despite* the legitimacy of this inductive projection and, consequently, despite believing that there are probably serious and well-confirmed presently unconceived alternatives to them. This latter avenue of response will be considered in depth in chapters 6 and 7.

11. Of course, as we will see over the course of the next several chapters, the case for the new induction will itself sometimes involve more specific eliminative arguments to the effect that the best explanation of the specific historical evidence available in a particular case is that a given scientist or scientific community simply failed to conceive of a given category of theoretical alternatives. But once again, these eliminative inferences concerning the interpretation and significance of the historical evidence will be of a kind for which we have no comparable general history of serious unconceived alternatives to cast doubt on their reliability.

3

Darwin and Pangenesis

The Search for the Material Basis of Generation and Heredity

On the origin of species Mr. Darwin has nothing, and is never likely to have anything, to say; but on the vastly-important subject of inheritance, and the transmission of peculiarities once acquired through successive generations, this work is a valuable storehouse of facts for curious students and practical breeders.

—from a review of Darwin's *The Variation of Animals and Plants Under Domestication* in the *Athenaeum*, Feb. 15, 1868

3.1 Preliminary Worries

To put it mildly, considerable room remains for skepticism regarding the sweeping historical thesis I have proposed. Did past scientists really fail to conceive of serious later alternatives to their theoretical accounts of the natural world altogether? Did they not rather conceive of and then dismiss such alternatives at least in broad outline or under broad categorical headings? And even if the relevant later alternatives were unconceived, were they really even serious at the time *given* the available evidence, not to mention such important differences as prevailing metaphysical presumptions and methodological standards?

I hope to confront such challenges in a fashion that will no doubt strike some readers as unpromising, if not perverse: by looking for evidence of the historical predicament I have claimed we occupy just where we might expect it to be hardest to find. I confess that I expect to find some sympathy for my general thesis among physicists and philosophers of physics: the revolutionary and counterintuitive character of such conceptual innovations as the electromagnetic field, general relativity, and virtually all things quantum mechanical has left many of those knowledgeable in the physical sciences with a healthy respect for presently unconceived possibilities and the dramatic ways in which the conception of the physical world embraced by our successors might (even probably will?) differ dramatically from our own. I will therefore focus my attention on the history of an area of scientific inquiry that seems at first glance far less naturally suited to the central claims of the new induction: the biological sciences. Not only is there a staunch tradition of scientific realism among biologists and philosophers of biology alike, but the history of the field is also one in which the challenges mentioned above seem particularly salient. If exploring the biological sciences does not simply kill off any enthusiasm we might have for the new induction or the problem of unconceived alternatives altogether, we might reasonably expect it to put us in a better position to know what to say about such challenges.

The specific area of biological science I will consider is the history of our theorizing about the phenomena of generation and inheritance. This will surely strike many readers as an unfamiliar description of any domain of biological inquiry at all, but it is nonetheless a fairly recent development that the study of how traits are passed on from parents to offspring (as in contemporary genetics) has become separated theoretically from the study of the formation and development of a new organism from germinal materials in the first place (as in contemporary embryology), or even that our theories concerning the processes of growth, reproduction, inheritance, development, and repair (healing and regeneration) have become sufficiently distinct to occupy different parts of the scientific landscape (see Dunn 1965 34, Bowler 1989 chap. 2 (esp. 23f), Gasking 1967 7–9, Maienschein 1986 esp. 101–103).[1] Not until the twentieth century did theorists clearly distinguish questions about how traits are passed on from parents and other ancestors to their offspring from questions about how a novel organism is generated (or regenerated) from germinal materials in the first place: since reproduction was thought of as a process in which parental organisms *manufactured* offspring from the materials of their own bodies, there simply was no sharp boundary or separation between questions about how this process might proceed and how traits of the parents might influence or be reflected in the resulting organism (see Bowler 1989 23–24). Until comparatively recently, then, the phenomena of inheritance, reproduction, development, growth, and repair were typically regarded as aspects of a single process, forming a single domain of theorizing and a single field of study (most commonly described simply as the study of 'generation'). Of course, even now we regard these fields as importantly related and mutually constraining: their disciplinary separation has been imposed more by the independent successes of the divergent theoretical and methodological tools of genetics and embryology than by some discovery that earlier theorists were wrong to regard the development of a new organism from germinal materials and its inheritance of characteristics from its ancestors as different aspects of a single continuous process.

Even with the historical landscape thus clarified, however, theories of generation and inheritance would seem at first glance to offer little in the way of promising raw material supporting either the new induction or the problem of unconceived alternatives. This is because the history of the field seems most broadly characterized by simple *alternation* between mechanistic and vitalist approaches, that is, between these approaches are attempts to explain some subset of the phenomena of growth, reproduction, inheritance, development, and repair by appeal to mechanical causal processes, on the one hand, and appeals (whether despairing or enthusiastic) to vitalistic forces or processes unique to the organic realm and irreducible to ordinary mechanical interactions, on the other (see Gasking 1967 passim, Bowler 1989 chap. 2). Thus, there is simply no question that later theories were typically anticipated in some *general* way or under some general description by earlier supporters and detractors of the same broad tradition (mechanical or vitalistic) in which they themselves can be located: for example, when nineteenth century teleomechanists denied that *any* purely mechanical process could produce the goal-driven phenomena of embryological development (see below), there is surely some sense in which contemporary molecular genetics (and its purely mechanical account of ontogeny) must be counted among the broad class of theories thereby preemptively rejected.

As this very example illustrates, however, details matter. More specifically, the failures of earlier theorists to conceive of or consider later accounts in any significant detail is crucial, because the profound differences between succeeding accounts in the same general tradition of biological thought are substantial enough to prevent later theories from being mere revisions, modifications, or further specifications of earlier ones: thus, if we are presently ignoring serious and well-confirmed competitors as different from our own theories as contemporary molecular biology is from equally 'mechanistic' eighteenth-century preformationism or nineteenth-century teleo-mechanism was from Aristotelian vitalism, the confirmational power of the eliminative support for current theories is undermined *whether or not* we have already captured these alternatives verbally under some broad categorical description like 'vitalism' or 'materialism.' And it will not be enough to undermine the new induction that earlier theorists managed to conceive of later alternatives in this abstract and general way. Once this point is appreciated, it is fairly easy to identify any number of important episodes in the history of our theorizing about generation and inheritance in which failures to conceive of or exhaust the space of serious alternative possibilities are clearly identifiable, even if we limit ourselves to the modern period of such theorizing (traditionally beginning with the publication of Harvey's *De Generatione Animalium* in 1651). For example, the eighteenth century witnessed a fundamental theoretical conflict between preformationists (like Malpighi, Haller, Bonnet, and Spallanzani) who believed that the parts of embryos were fully formed in miniature from the moment of conception, if not the moment of Creation, and so-called epigeneticists (like d'Holbach, Maupertuis, and Buffon) who believed that the parts of an embryo are produced sequentially in a developmental process. But it is an historical commonplace that without any sophisticated chemistry or grasp of molecular complexity and without the benefit of cell theory, *neither* group could form any concrete conception of *how* complex structures could form sequentially in the developing embryo by purely material processes (Roe 1981 chap. 1 (esp. p. 15), Bowler 1989 24–34; see also Gasking 1967 166). Instead, as Bowler notes, "[w]henever organic processes of this complexity were hinted at, they were taken as an excuse for qualifying the mechanical philosophy by hinting at hylozoism (the belief that matter itself is alive).... without an adequate conception of molecular complexity, it was only possible to explain so complex a process by attributing the properties of life to matter itself" (1989 32–34). Likewise, Spallanzani's famous experimental refutations of spontaneous generation (of living organisms from inorganic matter) could serve as "a vital line of support for pre-existing germs as far as Spallanzani himself was concerned" (Bowler 1989 30) only because he and other eighteenth-century preformationists regarded the existence of fully organized and developed miniature organisms encased in eggs or seeds as the *only* possible or plausible alternative to the spontaneous generation of organized, living beings from inert matter.[2] It was for this reason that they took preformationism to be supported or even simply established by the experimental demonstration that such spontaneous generation does not occur (see Bowler 1989 chap. 2, Mazzolini and Roe 1986 Introduction (esp. Section 2), Roe 1981 19). Spallanzani's *own* conclusion from his experiments showing that no organisms would generate within meticulously sealed and sterilized infusions was that "it is not possible to explain the birth of Animalcules by anything besides little eggs, or seeds, or preorganized bodies [*corpuscules préorganisés*], that I wish to call and will call by the generic name of *Germs*" (1776, original emphasis, my translation; quoted in Mazzolini and Roe 1986 49).

We might consider a more recent example in slightly greater detail. The well-deserved recent dominance of the mechanistic tradition in the life sciences has often encouraged an implausibly paradoxical historical image of German biology in the late eighteenth through the mid-nineteenth centuries: the period has sometimes been held to be characterized by dramatic empirical advances in accurate observation and description (especially in microscopic anatomy and descriptive embryology) achieved by thinkers who (influenced by the romantic *Naturphilosophen* movement) held completely confused, misguided, and unscientific (even mystical) vitalist beliefs about the natural processes they were observing. More recent historical work in the last several decades has put the lie to this unconvincing picture of scientists laboring in the grip of a conception of the workings of nature barely worthy of the name 'scientific' which fortunately did not prevent them from making dramatic empirical progress, or indeed, influence their scientific work in any important way. Timothy Lenoir has argued convincingly, for example, that the mystical developmentalism of the *Naturphilosophen* gave way among life scientists of Germany between 1790 and 1860 to a serious, concrete, successful, and undoubtedly scientific program of research in the life sciences (Lenoir 1981, 1982; see also Roe 1981 chap. 6, Gasking 1967 chap. 13). This "teleo-mechanist programme" of such thinkers as Johann Friedrich Blumenbach, Johann Christian Reil, Carl Friedrich Kielmeyer, Johann Friedrich Meckel, and Karl Ernst von Baer insisted on the need to posit the existence of vital or formative forces (*Lebenskraft, Bildungstrieb, Gestaltungkraft*) unlike those operative in the inorganic realm in order to account for organic phenomena like those of ontogeny, in which a developing embryo is invariably conducted to just one end state from any number of widely varying initial states or in which the activity of particular causal mechanisms seems to be directed towards the achievement of a particular goal or outcome.[3]

The teleo-mechanists insisted that such forces (along with the organic processes they directed) were neither reducible to nor explicable by ordinary inorganic chemical and electromagnetic interactions, affinities, or forces (notwithstanding, they argued, the successful inorganic synthesis of urea by Wöhler). But they nonetheless insisted with equal vigor that these forces must be a part of the ordinary material world, explicitly and categorically rejecting the mystical vitalism attributing the organization of organic entities to their direction by an immaterial soul that they held to be discernible in the writings of earlier thinkers like G. E. Stahl (probably fairly) and Caspar Friedrich Wolff (probably unfairly; see Roe 1981 chap. 4, esp. 109–110). Reil and Blumenbach defended the postulation of such forces by explicit analogy to Newtonian gravitation: that inert matter exerts and experiences an attractive force does not itself admit of further mechanical explanation, they pointed out, but the existence and properties of this force, like those of the vital forces responsible for organic phenomena, can nonetheless be known through its effects. Similar forces were ascribed a direct causal role in such physiological processes as contractility, irritability, and sensibility (see Larson 1979).

Perhaps the most impressive achievement offered by the teleo-mechanist program was its ability to provide a convincing causal explanation of a series of structural parallels whose depth and detail were becoming increasingly salient at this time. Not only was systematic taxonomy uncovering a remarkable recurrence of the same basic anatomical and physiological plans across widely varying groups of organisms, but there was also emerging a striking and undeniable correspondence between the succession of forms of organismic diversity as they first appeared in the Earth's fossil record and the succession of stages undergone by present organisms during the course of their embryological development in both normal and teratological (monstrous) cases. Teleo-mechanists had an explanation close to hand for this unexpected and otherwise astounding correspondence. As Kielmeyer noted,

> The idea of a close relationship between the developmental history of the earth and the series of organized bodies, in which each can be used interchangeably to illuminate the other, appears to me to be worthy of praise. The reason is this: Because I consider the force by means of which the *series* of organized forms has been brought forth on the earth to be in its essence and the laws of its manifestation *identical* with the force by means of which the series of developmental stages in each *individual* are produced, which are *similar* to those in the series of organized bodies. . . . (Kielmeyer 1793–4, original emphasis; cited in Lenoir 1981 314; see also Coleman 1973)

Because teleo-mechanists held the same developmental or formative vital force to be operating in producing the various forms of any historically related lineage of organisms, it seemed perfectly natural that the embryological development of any individual organism would involve a recapitulation of the series of historically preceding forms produced by the force in question:

> Since the *formative force*, whatever it is, has less energetic impulse [in lower animals] than in higher animals, the organs run through only a part of the transformations that they undergo in superior creatures. From this it follows that they offer to us, in a permanent manner, the organic configurations that are only transitory in the embryo of man and the higher vertebrates. (Serres 1830; cited in Gould 1977 49)

Thus teleo-mechanists did not simply invoke the existence of a vital force wherever they found phenomena they could not otherwise explain, but instead postulated a small group of specific formative forces responsible for the structured development of living organisms, corresponding roughly to the main divisions of the organismic world as they saw them. Each of these forces directed and constrained the mechanical pathways of growth and development in a single distinct and definite series of organisms, and was passed from generation to generation through the germinal materials. A similarly recapitulationist conception of the relationship between ontogeny and phylogeny lead Meckel in 1811 to predict the appearance of gill slits in human embryonic development, a prediction subsequently confirmed (though interpreted differently) by Rathke and von Baer in the late 1820s (see Gould 1977 46).

Thus, the postulation of formative or developmental forces figured in widely respected and genuinely (even predictively) successful scientific practices of the first half of the nineteenth century. Even when Von Baer famously argued *against* the genuine recapitulationism of Kielmeyer and Meckel (that is, against the expression of *adult* stages of ancestral organisms during embryonic development), he insisted nonetheless that the existence of specific vital forces in different grades of development were the key to explaining the striking relationships between the corresponding *embryonic* stages of various organisms and specific organs. Von Baer suggested that similarities occurred not because later organisms recapitulate earlier ones during embryonic development, but instead because related organisms share aspects of a single developmental and regulative 'type':

> Since, in fact, the embryo becomes gradually perfected by progressive histological and morphological differentiation, it must in *this respect* have the more resemblance to less perfect animals the younger it is. Furthermore, the different forms of animals are sometimes more, sometimes less remote from the principal type. The type itself never exists pure, but only under certain modifications. But it seems absolutely necessary that those forms in which animality is most highly developed should be furthest removed from the fundamental type.... The Worms, the Myriapoda, have an evenly annulated body, and are nearer the type than the Butterfly. If then the law be true, that in the course of development the principal type appears first, and subsequently its modifications, the young Butterfly must be more similar to the perfect Scolopendra and even to the perfect Worm, than conversely the young Scolopendra or the young Worm to the perfect Butterfly.... The same thing is obvious in Vertebrata. Fishes are less distinct from the fundamental type than Mammalia, and especially than Man with his great brain. It is therefore very natural that the Mammalian embryo should be more similar to the Fish than the embryo of the Fish to the Mammalian. (Von Baer 1828 [1951] 398, original emphasis)

Von Baer argued that embryological investigation revealed in the animal kingdom precisely four such regulative 'types,' corresponding roughly to Cuvier's division of the animal kingdom into four *embranchements* (Radiata, Articulata, Mollusca, Vertebrata). Thus, teleo-mechanists could not have failed to see the evidence as converging from a variety of independent sources to confirm the existence of specific vital forces common to distinct groups of organisms: Cuvier's taxonomic division was grounded in systematic anatomy and physiology, but seemed to reveal the detailed footprints of precisely the same specific constructive or formative vital forces suggested by the independent evidence from experimental embryology and from the fossil record.

Perhaps most important of all for the purposes of the new induction, however, is the fact that such theorists as Reil, Blumenbach, Kielmeyer, Weber and von Baer argued explicitly that the directive or teleological character of many organic causal processes absolutely *required* the postulation of such special formative or developmental forces and that there was *simply no other way* to account for the many organic phenomena in which an end product or goal, or a functional relationship not yet realized, seemed to direct specific causal pathways of interaction and development:

> From such evidence it must be concluded that the formative force in organic bodies is able to give the larger parts their determinate position and shape even when the smaller parts from which they are constructed have a variable shape and positional arrangement: and consequently the shape of the whole organs cannot depend upon the affinity of the smallest parts for one another, which they exercise in virtue of their inherent qualities, and which causes them to assume determinate positions with respect to one another, as appears to be the case in crystallization. On the contrary, the formative activity must be determined by laws, which are connected with the interrelationships with respect to size, form and position of the larger parts of the organized body; *i.e.* independently of the interrelationships of the smallest parts.
>
> Complex crystals are constructed from individual *parts*; organisms, however, are constructed from the *whole*. (Weber 1830; cited in Lenoir 1981 318)

Early teleo-mechanists repeatedly drew this contrast between the operation of their formative vital forces and processes like crystallization, in which complex structure emerges simply as a consequence of aggregated local physical and chemical interactions. That is, teleo-mechanists insisted that simply aggregating such local physical and chemical interactions could not *possibly* explain the goal-directed processes of ontogeny and physiology. As Lenoir points out, Kielmeyer defended his own teleological vital force, the *Lebenskraft*, by insisting:

> that the inorganic forces of mechanics and chemistry by themselves could only result in the formation of chemical-physical 'aggregates'. They could not lead to the determination of a functional, integrated, whole organism, where each part exists and only has being with reference to the whole of which it is a part. (1981 317)

Von Baer defended a similar claim with particular attention to the phenomena of embryonic development, pointing out that during such development organisms are led from a wide variety of extremely variable initial states to just a single outcome:

> it must be concluded that... every variation, as far as it is possible, is conducted back to the norm. For this, however, it is clear that each temporary condition by itself cannot determine the future state of beings; on the contrary the more general and higher relations must dominate the process. (1828; cited in Lenoir 1981 321)

In Von Baer's (and later Rathke's) hands, this conception of embryological development under the influence of formative forces grounded predictions about and explanations of the origin of various organisms' genuinely homologous organs from identical areas of the germ layers, as well as Von Baer's famous laws of development and his predictions about the invariable course of development in all particular cases. Von Baer ultimately went so far as to offer what he regarded as convincing empirical evidence for the physical location of such a directive or formative vital force, the *Gestalungskraft*, at the center of the mammalian ovum (his own most famous observational discovery).

In short, vital teleo-mechanists of the nineteenth century argued explicitly and repeatedly that the developmental or formative forces they (sometimes reluctantly) recognized were the only possible way to explain a wide variety of organic processes. The postulation of such vital forces was neither an embrace of mysticism, as it has sometimes been portrayed, nor was it divorced from the very real successes (even the predictive successes) of German biology in the late eighteenth and early-to-mid nineteenth centuries. But most importantly, the insistence of teleo-mechanists that the postulation of such vital forces was simply *unavoidable* in understanding apparently goal-directed organic phenomena is a clear and convincing reflection of their failure to conceive of or consider the sort of sophisticated materialistic explanations of organic processes that would come to dominate the study of the life sciences in the generations that followed them.

Perhaps we have by this point responded to the first and second challenges with which we began. The history of biological theorizing about generation and inheritance does indeed seem to exhibit failures to conceive of alternative theoretical possibilities sufficiently serious as to ultimately attract groups of sincere scientific defenders. We have also seen that the new induction is not threatened by the fact that past theorists have sometimes anticipated later alternatives in some general form or under some general verbal description, for their failure to consider such alternatives in detail suffices to undermine the confirmational credentials of the eliminative inferences that ignore them. But for some readers the exploration of nineteenth-century teleo-mechanism will undoubtedly arouse with a vengeance the third and perhaps most serious of the challenges we raised at the outset: were the later alternatives unconceived by earlier practitioners really even serious ones *at the time*, given profound differences in available evidence, metaphysical presuppositions about nature, and methodological assumptions about its investigation?

We might note first of all that this way of responding to the new induction carries its weight only if we assume that *today's* methodological assumptions and metaphysical presuppositions will (probably and/or approximately) persist indefinitely into the future as inquiry proceeds and that there are no dramatic evidential surprises waiting around the bend for *us*: otherwise, even if past practitioners did manage to exhaust the space of plausible or serious alternatives *by the standards of the day*, this offers no reassurance to the eliminative support for contemporary theories, for we will recognize ourselves as presently exhausting a space of possibilities that will itself not remain stable in the face of expected changes in our own evidence, metaphysics, and methodology.

But of course this response threatens to simply push the historical challenge to any claim of privilege for contemporary science back to the level of metaphysics, methodology, and evidence and invite us to start over, so we should try not to rely on it if we can avoid doing so. Accordingly, I will instead try to show that a considerably deeper descent into the details of the historical record establishes that our efforts to theorize about inheritance and generation continued to be plagued by the problem of unconceived alternatives long after we came to embrace substantive evidential, metaphysical, and methodological constraints essentially continuous with those of the present day. I will undertake this descent in the remainder of this chapter and over the course of the next two. My goal will be to show that even after mid-to-late nineteenth-century theorists sought to develop a material, particulate account of generation and inheritance under roughly the same constraints, directed at roughly the same phenomena, and articulated along roughly the same general metaphysical principles as contemporary genetics and embryology, they repeatedly failed to conceive of scientifically serious and well-confirmed alternatives to their own proposals. Such seminal thinkers as Charles Darwin, Francis Galton, and August Weismann offered fully materialist and mechanical theories of inheritance directed largely (though as we will see, not exclusively) toward the same general phenomena for which we take contemporary genetic and embryological theories to be explanatorily responsible. Nonetheless, I will suggest that the accounts offered by each of these important theorists and the arguments used to advance them reflect profound failures to conceive of entire *classes* of theoretical alternatives whose representative members would be actually embraced by later theorists and scientific communities. And I will suggest that each of the major conceptual advances we now recognize in the search for a material substrate of inheritance represents the recognition of an entire suite of theoretical possibilities that were scientifically serious even by the standards of the day despite being unconceived and therefore unconsidered by theorists at the time.

3.2 Pangenesis: Darwin's "Mad Dream" and "Beloved Child"

From the middle to the end of the nineteenth century, interest in identifying some material basis for the transmission of characteristics from parents to offspring gained dramatic momentum from such converging influences as increasingly detailed microscopic observations, the development of cell theory, and advances in experimental hybridization. But each of these developments was in turn prompted at least in part by the publication (in 1859) of Charles Darwin's *Origin of Species* and the importance thereby conferred on questions about the mechanism of evolution and, consequently, about the sources of variation in nature (see Gayon 1998 chap.1; Dunn 1965 34; Olby 1985 chap. 3; Robinson 1979 xiii, 3; Cowan 1985 chap. 5; Bowler 1989 46; Gasking 1967 161; Geison 1969 375, 385–386). Though various kinds of hereditary particles had been proposed by Buffon, Diderot, Maupertuis and others in the period before Darwin, the idea of living or material units or particles[4] as the substrate of inheritance that is developmentally continuous with our own is usually traced back to the "physiological units" introduced by Herbert Spencer in his *Principles of Biology* (1864) and to the "gemmules" of Darwin's own theory of pangenesis, first proposed in his *Variation of Animals and Plants Under Domestication* in 1868 (hereafter *Variation* or VAP[5]) Although Spencer's version of the general idea was published four years earlier, Darwin's account appears not to have been based upon it, as he had by that point been developing the theory of pangenesis for some twenty-odd years.[6] Indeed, Spencer's conception of physiological units (first introduced in his section on "Waste and Repair") offered much vague talk of their "proclivities" and "special polarities"; it was not without some justice that Darwin ultimately claimed of the *Principles* that "each suggestion, to be of real value to science, would require years of work" (Darwin to Hooker, June 30 [1866] in *More Letters of Charles Darwin* (hereafter MLD) v. 2 235) and Darwin was relieved to hear Spencer himself say there was a profound difference between their respective views.[7] In any case, Darwin's much more concrete and more clearly mechanistic hypothesis of pangenesis would exercise a greater influence on subsequent theorizing about generation and inheritance, and it was Darwin's theory that later theorists of generation would feel obliged to confront and discuss, even if only to abuse it.

Famously, of course, during nearly the entire period of theoretical development we will consider, Mendel's discovery of the ratios with which parental traits reappear in subsequent generations of hybrid offspring and their suggestive implications concerning the mechanism of heredity (published in the journal of the Brno Natural History Society in 1866) lay largely unknown and unappreciated in libraries across Europe, including those of the Royal Society and the Linnean Society in Great Britain (see Olby 1985 103). The foundational significance that would ultimately be attributed to these experimental results in suggesting a general hereditary mechanism of paired dominant and recessive material particles governing the expression of particular traits went unrecognized by the celebrated botanist Carl Nägeli (Mendel's only influential correspondent in the scientific world), who replied to them only by encouraging Mendel to try to repeat the experiments with *Hieracium* (Hawkweed), an experimental organism of special interest to Nägeli himself, but one that does not exhibit the characteristic Mendelian segregation of characters (an effect now attributed to apomixis). Only after suggestive microscopic discoveries concerning chromosomal behavior would the Mendelian ratios (and Mendel himself) be independently "rediscovered" by Hugo de Vries, Erich von Tschermak and Carl Correns. This case constitutes a particularly interesting source of support for the problem of unconceived alternatives, as it offers an especially clear testament to the inability to even recognize a particular unconceived alternative theoretical explanation for which the data, to modern eyes, seem to cry out. This was certainly true for Nägeli, a leading nineteenth-century theorist of generation and inheritance, and it has been argued that Darwin could have fared no better in recognizing the significance of Mendel's results had they been presented to him (Bowler 1989 56, see also Olby 1985 52). Such claims are not unrelated to the influential historiographical thesis (see Olby 1979, 1985; Callender 1988; Bowler 1989 chap. 5) that the "rediscovery" of a theoretical model of inheritance proposed by Mendel and subsequently neglected is *itself* simply a myth, for Mendel's aim was to show that new species arose by hybridization rather than transmutation (a thesis for which *Hieracium* provided even better evidence than his original work with *Pisum*) and *Mendel himself neither recognized nor proposed the mechanism of hereditary transmission that de Vries, Tschermak, Correns, and biologists of today find leaping off the pages of his experimental work*!

In any case, Darwin was certainly not influenced by Mendel's discoveries during the several decades he spent gathering information about and reflecting on the phenomena of inheritance, generation, growth, and development before publishing his own material and mechanical theory of inheritance and generation in 1868 (Robinson 1979 4–5). Furthermore, as any number of commentators have pointed out, Darwin did not share our view of heredity and variation as complementary aspects of a single process, but instead subscribed without substantial reflection to a longstanding view of these as antagonistic forces or principles operating in opposition to one another (e.g. Churchill 1987 343–345; Bowler 1989 25, 68; Hodge 1989 272–273, 277; see also Gayon 1998 chap.1). Darwin thus came to suggest that variations between parents and offspring were anomalous incidents, produced largely if not exclusively by changes or irregularities in the "conditions of life" and taking place against a broad background of inherited characteristics: he suggests that "we may on the whole conclude that inheritance is the rule, and noninheritance the anomaly" (VAP v. 2 454) and that the "proper function" of reproductive systems is "transmitting truly the characters of the parents to the offspring" (VAP v. 2 453).[8] Against this theoretical background, here is Darwin's own description of pangenesis as it appeared in the second (1874) edition of the *Variation*:

> It is universally admitted that the cells or units of the body increase by self-division or proliferation, retaining the same nature, and that they ultimately become converted into the various tissues and substances of the body. But besides this means of increase I assume that the units throw off minute granules which are dispersed throughout the whole system; that these, when supplied with proper nutriment, multiply by self-division, and are ultimately developed into units like those from which they were originally derived. These granules may be called gemmules. They are collected from all parts of the system to constitute the sexual elements, and their development in the next generation forms a

> new being; but they are likewise capable of transmission in a dormant state to future generations and may then be developed. Their development depends on their union with other partially developed or nascent cells which precede them in the regular course of growth. . . . Gemmules are supposed to be thrown off by every unit, not only during the adult state, but during each stage of development of every organism; but not necessarily during the continued existence of the same unit. Lastly, I assume that the gemmules in their dormant state have a mutual affinity for each other, leading to their aggregation into buds or into the sexual elements. Hence, it is not the reproductive organs or buds which generate new organisms, but the units of which each individual is composed. These assumptions constitute the provisional hypothesis, which I have called Pangenesis. (VAP v. 2 457)

Darwin writes that it is the evident relation between "large classes of facts, such as those bearing on bud variation, the various forms of inheritance, the causes and laws of variation" and "the several modes of reproduction" which have "led, or rather forced" him to form a view connecting them (VAP v. 2 432). And he offers a characteristically exhaustive list of phenomena for which he suggests pangenesis alone can account:

> How it is possible for a character possessed by some remote ancestor suddenly to reappear in the offspring; how the effects of increased or decreased use of a limb can be transmitted to the child; how the male sexual element can act not solely on the ovules, but occasionally on the mother-form [under this heading Darwin also later includes its effect on the offspring of later matings]; how a hybrid can be produced by the union of the cellular tissue of two plants independently of the organs of generation; how a limb can be reproduced on the exact line of amputation, with neither too much nor too little added; how the same organism may be produced by such widely different processes, as budding and true seminal generation; and lastly, how of two allied forms, one passes in the course of its development through the most complex metamorphoses, and the other does not do so, though when mature both are alike in every detail of structure. (VAP v. 2 432–433)

As Darwin saw it, the central idea capable of explaining each of these disparate phenomena and of unifying them all was that "an organism does not generate its kind as a whole but each separate unit generates its kind" (VAP v. 2 490). More fully, "every separate part of the whole organization reproduces itself. So that ovules, spermatozoa, and pollen-grains—the fertilized egg or seed, as well as buds—include and consist of a multitude of germs thrown off from each separate part or unit" (VAP v. 2 433). As Hodge (1985, 1989) rightly emphasizes, this proposal that all forms of generation were continuous and unified by a single mechanism represented a radical departure from Darwin's earlier views, which had assumed a fundamental distinction between sexual and asexual forms of generation. He grants that his account is "merely a provisional hypothesis or speculation" which might involve incompleteness or error, but insists nonetheless that "until a better one be advanced, it will serve to bring together a multitude of facts which are at present left disconnected by any efficient cause" (VAP v. 2 433).

An important source of Darwin's insistence that these phenomena of generation and inheritance had yet to be connected by "any efficient cause" and that pangenesis alone provided an explanation of them was his refusal to regard appeals to vitalistic powers or potentials as offering any genuinely explanatory *alternative* to pangenesis at all. In the *Variation* he argues that such talk of potentialities and powers should be understood *in terms of* the central theoretical mechanism postulated by pangenesis: "It has often been said by naturalists that each cell of a plant has the potential capacity of reproducing the whole plant; but it has this power only in virtue of containing gemmules derived from every part" (VAP v. 2 490). Similar sentiments appear in the Author's Preface (dated March 28, 1868) to the first American edition of the *Variation*:

> I venture to call the reader's attention to the chapter on Pangenesis. The view there propounded is simply hypothetical, but it has appeared to me . . . to be no small gain to seize on a material bond, by which the various forms of reproduction inheritance, development, etc. can be connected together. We thus get rid of such vague terms as spermatic force, the vivification of the ovule, sexual potentiality, and the diffusion of mysterious essences or properties from either parent, or from both, to the child." (v. 1 i)

But Darwin's insistence that vitalistic appeals to powers or potentials offered no genuine explanatory competitor to pangenesis is perhaps most eloquently expressed in a letter to Hooker written just a month after the publication of the *Variation* in 1868:

> When you [Hooker] or Huxley say that a single cell of a plant, or the stump of an amputated limb, have the "potentiality" of reproducing the whole—or "diffuse an influence," these words give me no positive idea—but when it is said that the cells of a plant, or stump, include atoms derived from every other cell of the whole organism and capable of development, I gain a distinct idea. But this idea would not be worth a rush, if it applied to one case alone; but it seems to me to apply to all the forms of reproduction—inheritance—metamorphosis—to the abnormal transposition of organs—to the direct action of the male element on the mother plant, &c. Therefore I fully believe that each cell does *actually* throw off an atom or gemmule of its contents; —but whether or not, this hypothesis serves as a useful connecting link for various grand classes of physiological facts, which at present stand absolutely isolated. (Darwin to Hooker, February 28 [1868] in LLD v. 2 264)

Besides illustrating his reasons for thinking that vitalistic appeals offered at best an intolerably vague description of the sort of mechanism posited by pangenesis itself, this letter also offers clear testimony that Darwin himself conceived of the support for his hypothesis as eliminative in character: the central virtue he claims for pangenesis is that it alone offers a "positive" or "distinct" idea capable of explaining and unifying a wide variety of the phenomena of heredity and generation "which at present stand absolutely isolated." Furthermore, Darwin here reports that this eliminative foundation was sufficient to lead him to "fully believe" in the literal truth of at least the theory's central claim that each cell does indeed throw off gemmules destined to develop into cells like the one from which they arose.

But how can we know that Darwin really failed to conceive of possible mechanistic alternatives to pangenesis at all, rather than simply finding sufficient fault to dismiss them out of hand as unacceptable, as he seems to have treated Hooker's conception of vitalistic powers? While we will find later theorists revealingly argue that particular aspects of their own theories are simply entailed or required by any hypothesis of physiological units of inheritance (Galton), or by the empirical phenomena themselves (Weismann), Darwin never suggests that the phenomena of inheritance, growth, development, reproduction, and repair could not *possibly* be otherwise explained. Instead he offers explicit and repeated assurances (even in the title of the chapter in which it is proposed) that his hypothesis is "provisional" and tentative, apparently in response to what seems to have been a skeptical reaction by Huxley to the pangenesis manuscript of 1865 (see Olby 1963; Robinson 1979 16).

Nonetheless, despite this characteristic caution with which Darwin presented to the world the theory he told Gray "will be called a mad dream" (October 16 [1867] in LLD v. 2 256), his private correspondence offers convincing evidence that he really did fail to conceive of relevant alternatives: besides remarking (in the passages noted above) that the known phenomena of heredity and generation are otherwise "absolutely isolated" and "disconnected by any efficient cause," Darwin repeatedly *tells* his correspondents that pangenesis is the first and only theory he has conceived of that can account for them. In asking Huxley to review his manuscript of the proposed chapter on pangenesis in the first place he writes as follows:

> in my next book [VAP] I shall publish long chapters on bud- and seminal-variation, on inheritance, reversion, effects of use and disuse, &c. I have also for many years speculated on the different forms of reproduction. Hence it has come to be a passion with me to try to connect all such facts by some sort of hypothesis. The MS. which I wish to send you gives such a hypothesis; it is a very rash and crude hypothesis, yet it has been a considerable relief to my mind, and I can hang on it a good many groups of facts. (Darwin to Huxley, May 27 [1865?] in LLD v. 2 227–228)

He writes to Hooker that "though I can see how fearfully imperfect, even in mere conjectural conclusions, it is; yet it has been an infinite satisfaction to me somehow to connect the various large groups of facts, which I have long considered, by an intelligible thread" (Darwin to Hooker, November 17 [1867] in LLD v. 2 257). He takes himself to echo Wallace's own feelings in saying "that it is a relief to have some feasible explanation of the various facts, which can be given up as soon as any better hypothesis is found. It is certainly an immense relief to my mind; for I have been stumbling over the subject for years, dimly seeing that some relation existed between the various classes of facts" (Darwin to Wallace, February 27 [1868] in LLD v. 2 262 and in MLD v. 1 301). To Hooker he quotes Wallace[9] as saying "It is a *positive comfort* to me to have any feasible explanation of a difficulty that has always been haunting me, and I shall never be able to give it up till a better one supplies its place, and that I think hardly possible, &c.," adding that Wallace's words "express my sentiments exactly and fully: though perhaps I feel the relief extra strongly from having during many years vainly attempted to form some hypothesis" (Darwin to Hooker, February 28 [1868] in LLD v. 2 264 original emphasis). He tells G. Bentham that "to my mind the idea has been an immense relief, as I could not endure to keep so many large classes of facts all floating loose in my mind without some thread of connection to tie them together in a tangible method" (Darwin to G. Bentham, April 22 [1868] in MLD v. 2 371). And he writes to Fritz Müller that "I find it a great relief to have some definite, though hypothetical view, when I reflect on the wonderful transformations of animals, the regrowth of parts, and especially the direct action of pollen on the mother-form, &c." (Darwin to Müller, June 3 [1868] in MLD v. 2 82). Thus we seem faced with a wealth of occasions on which Darwin simply *reported* that pangenesis was the only hypothesis he knew of or had conceived of that could explain the diverse phenomena of generation and inheritance that had attracted his attention.[10] If Darwin did consider alternative possibilities or proposals for a mechanistic account of heredity and generation, he worked hard to keep us from knowing about it, for (in stark contrast to his treatment of vitalistic powers) none of these various reflections, assurances, or confessions show any evidence of entertaining and dismissing such alternatives; instead he repeatedly insists that the central mechanism posited by pangenesis is the lone contender.[11]

Given Darwin's apparent inability to conceive of any alternative to the basic strategy pangenesis offered for explaining the phenomena of heredity and generation, perhaps it is unsurprising that in his private correspondence Darwin was much less circumspect about the prospects for pangenesis and much more confident that his "beloved child" (Darwin to Hooker, February 3 [1868] in LLD v.2 258), "an infant cherished by few as yet, except his tender parent, but which will live a long life" (Darwin to Gray, May 8 [1868] in LLD v. 2 266), would ultimately win the day. To Huxley he writes that he is "becoming convinced that some such view will have to be adopted" (Darwin to Huxley, May 30 [1865] in Darwin (2002)), to Gray that he thinks it "contains a great truth" (Darwin to Gray, October 16 [1867] in LLD v. 2 256), to F. Hildebrand that he believes it "will ultimately be accepted" (Darwin to Hildebrand, January 5 [1868] in MLD v. 1 285) and to Müller that "Pangenesis will turn out true someday!" (Darwin to Müller, May 12 [1870] in MLD v. 2 359). To William Ogle he writes, "I advance the views merely as a provisional hypothesis, but with the secret expectation that sooner or later some such view will have to be admitted" (Darwin to Ogle, March 6 [1868] in LLD v. 2 265) and to J. J. Weir that "I fully believe pangenesis will have its successful day" (Darwin to Weir, March 17 [1870] in MLD v. 1 320). In an unpublished letter of July 14, 1868, Darwin advises Hooker not to touch on pangenesis in an upcoming address in light of the many luminary figures opposed to the theory, but nonetheless reports that "my conviction is unshaken that it will hereafter be looked at as the best hypothesis of generation, inheritance [and] development." And a later unpublished letter to J. V. Carus offers the similar view that "after mature reflection I believe that physiologists will some day be compelled to admit some such doctrine" (October 19 1868).[12] It seems hard to explain this assurance that pangenesis would ultimately be embraced unless we assume that its source lies in what Darwin elsewhere frankly admits: that he can think of no other mechanistic hypothesis that can account for the phenomena.

Moreover, Darwin explicitly links his confidence that pangenesis will triumph or reappear with his inability to conceive of or identify any alternative explanation for what he regarded as the central phenomena of heredity. After receiving Huxley's apparently sharp criticism of the pangenesis manuscript of 1865 he writes, "I do not doubt your judgment is perfectly just, and I will try to persuade myself not to publish. The whole affair is much too speculative; yet I think some such view will have to be adopted, when I call to mind such facts as the inherited effects of use and disuse, &c." (Darwin to Huxley, July 12 [1865?] in LLD v. 2 228). And to Hooker, Darwin again explicitly links his confidence that pangenesis will reappear with his inability to conceive of or identify any alternative explanation for the variety of hereditary phenomena for which he thinks pangenesis alone accounts:

> You will think me very self-sufficient, when I declare that I feel *sure* if Pangenesis is now stillborn it will, thank God, at some future time reappear, begotten by some other father, and christened by some other name.
>
> Have you ever met with any tangible and clear view of what takes place in generation, whether by seeds or buds, or how a long-lost character can possibly reappear; or how the male element can possibly affect the mother plant, or the mother animal, so that her future progeny are affected? Now all these points and many others are connected together, whether truely or falsely is another question,[13] by Pangenesis. (Darwin to Hooker, February 23 [1868] in LLD v. 2 261 original emphasis)

If we eschew the benefits of scientific hindsight and project ourselves into Darwin's position, it is easy to sympathize with his sense that pangenesis (or some close relative) represented the only *possible* mechanical explanation of the phenomena of generation and inheritance that had so captured his interest: after all, how could features of offspring so accurately reflect so many diverse peculiarities of their parents (no matter which of several different methods of reproduction gave rise to them) unless each of the parent's tissues, organs, and other physical features causally contributes to or serves as a physical template for the formation of the corresponding part of the bodies of its several offspring? Little wonder, then, that Darwin wrote to Müller that "It often appears to me almost certain that the characters of the parents are 'photographed' on the child, only by means of material atoms derived from each cell in both parents, and developed in the child" (Darwin to Müller, June 3 [1868] in MLD v. 2 82).

But once the question and Darwin's answer are phrased in this way, it seems easy in retrospect to articulate at least one broad class of serious theoretical alternative possibilities that seems to have escaped his consideration completely: parents and offspring might share salient characteristics not because the parents' tissues or other physical features themselves contribute materially or even causally to the formation of those of the offspring but instead because *both* sets of tissues, organs and features (with their shared peculiarities) are produced by *shared germinal materials*, of which identical or systematically related versions are invariably passed from parents to offspring. That is, the tissues of the offspring (produced by whatever intervening mechanism) might recapitulate salient features of the parent's not because the latter serve as causes of the former, but because they share a *common cause* in the hereditary materials found in a shared germ line ultimately producing them both.

Note that this suggestion does not require us to Whiggishly dismiss the full range of phenomena Darwin invoked pangenesis to explain and focus instead on just those unified and accounted for by more recent theories of inheritance: this is because the explanatory promise held out by pangenesis for the phenomena of heredity and generation holding Darwin's interest survives a shift from pangenesis' conception of hereditary particles as links in a causal chain (from the traits and developed tissues of the parent to those of the offspring) to the alternative idea of a shared germinal source of such particles serving as a common cause of traits and tissues in both parent and offspring. That is to say, if we allow that the processes of generation, inheritance, growth, development, and repair are mediated by hereditary particles distributed throughout the body but suppose that the source of such particles is a continuous germ line that can be passed in a variety of ways from parent to offspring rather than the developed tissues of the parent organism itself, we may still *retain* the explanations of hereditary phenomena that were the centerpiece of Darwin's case for pangenesis. These include Darwin's pangenetical explanations (VAP v. 2 467–488) of reversion, of bud-variation, of graft-hybrids, of parthenogenesis, of the development of complex tissues, of the processes of repair (and their precision), of the continuity between various forms of reproduction, of the possibility of producing identical organisms by both budding and seminal generation and with or without complex metamorphoses, and even of phenomena whose existence Darwin accepted but that we deny, like the direct influence of the "male sexual element" on the tissues of the mother plant (later called xenia or metaxenia) and on later progeny of the same female animal by different males (telegony).[14]

Perhaps most importantly of all, such a proposal would not have required Darwin to give up his famous commitment (especially late in life) to the inheritance of acquired characters,[15] because we need not suppose the germ line to be *isolated* in order to have the fundamental mechanical structure that Darwin fails to consider. We might suppose, for instance, that the germinal materials passed on to the offspring can themselves be affected by "mutilations and... accidents, especially or perhaps exclusively when followed by disease.... the evil effects of the long-continued exposure of the parent to injurious conditions.... the effects of the use and disuse of parts, and of mental habits" and "[p]eriodical habits" (VAP v. 2 70–71) without thereby giving up the idea that shared peculiarities of parent and child are quite *generally* effects of a common cause rather than links in a causal chain. That is, we might simply accept that the conditions in which the inheritance of acquired characters was supposed to occur were just those in which activities or events affecting the parent's body can exercise some influence on the shared germinal source of hereditary particles passed on to the offspring: we might even propose, as Galton would later in connection with his own common-cause account of hereditary structure (see below), a separate, gemmule-mediated mechanism to account for the phenomenon wherever (or *if* ever, as Galton would insist) its existence could be conclusively established. Indeed, this suggestion seems parallel to Darwin's own treatment of distant reversion: he accounts for the possibility by suggesting that gemmules will sometimes lay dormant for generations before developing (often triggered, he suggests, by hybridization or by changes in the "conditions of life"; VAP v. 2 455, 486), but he has almost nothing in the way of a substantive account to offer (see VAP v. 2 357) of why or the mechanism whereby they do so.[16] Reversion and the inheritance of acquired characters were perhaps the two most important puzzles about heredity for which Darwin hoped to account (see Geison 1969 388–391, 410; see also Endersby 2003 78–80), but it seems to involve no less of an explanatory lacuna to suggest that 'sometimes events during life can affect a shared germinal source of characteristics that is passed on to subsequent offspring' than it does to say of distant reversion simply that 'sometimes gemmules can lay dormant for generations before being developed.'

Furthermore, much of Darwin's own explanation of the inheritance of acquired characters can be preserved even on the assumption that shared characteristics of parents and offspring are effects of a common cause rather than links in a causal chain. The cases of the inheritance of acquired characters that Darwin regarded as most convincing were those in which the mutilation or amputation of a part of the parent was accompanied or followed by disease, rather than simply repeated for generations. His explanation of this difference was that "all the gemmules of the mutilated or amputated part are gradually attracted to the diseased surface during the reparative process, and are there destroyed by the morbid action" (VAP v. 2 484). And we can certainly retain this account of the *difference* between mutilations or amputations of diseased versus nondiseased tissues if we suppose that the constant morbid action preferentially depletes gemmules from a shared germinal source rather than from a supply already thrown off by the part in question before its amputation. Indeed, Darwin's explanation somewhat implausibly requires that removing the ultimate source of further gemmules (i.e. the amputated tissue or structure) in cases unaccompanied by disease has no effect on their later availability for reproduction, so the suggestion that morbid depletion grounds the difference in hereditary consequences between diseased and undiseased cases might even seem to fit rather *better* with the idea of a shared germinal source than with a causal chain from parental traits or tissues to those of the offspring in the first place.

Moreover, even if I am wrong to think that Darwin could have simultaneously embraced both the inheritance of acquired characteristics and a common-cause alternative to the structure of inheritance proposed in pangenesis, it would follow only that those cases of the inheritance of acquired characters of which Darwin was confident would have to count as empirical anomalies for any competing common-cause account of inheritance. But inheriting an anomalous phenomenon of this sort would not automatically disqualify the common-cause hypothesis as a serious contender to pangenesis for explaining the bulk of phenomena that concerned Darwin. He certainly recognized and tolerated any number of anomalies for pangenesis itself: in the *Variation*, for example, Darwin notes that pangenesis cannot explain why gemmules do not spread from bud to bud in plants (v. 2 462) and that it has no explanation for a number of differences in tendencies to reversion between plants propagated from buds rather than seeds (v. 2 480–481). His private correspondence also recognizes empirical anomalies for pangenesis, as when he writes to Hooker that "even Pan.[genesis] won't explain" the selective impotence of pollen when contacting ovules of same plant (May 21, 1868, in MLD v. 1 302). Similarly, the May 25, 1871, issue of *Nature* published a letter by A. C. Ranyard objecting to pangenesis on the ground that in graft hybrids, the "sexual elements produced by the scion" have not been shown to be affected by the stock, annotated in Darwin's copy simply as "The best objection yet raised" (MLD v. 1 302).

Finally, although belief in the inheritance of acquired characteristics was quite widespread among biologists at the time Darwin wrote (see Cowan 1985 62–63), the very existence of this phenomenon remained a disputed and controversial empirical matter even at this time. Perhaps the most influential support for the phenomenon came from famous experiments on guinea-pigs by the physiologist Brown-Séquard (see VAP v. 2 483; Robinson 1979 22; Cowan 1985 63–64), but Geison notes that "opinion was divided among influential nineteenth-century authors," as James Cowles Prichard, William Lawrence, and Joseph Hooker, for example, seem to have denied that the phenomenon occurred (1969 379n).[17] Not only was Darwin aware of this resistance to the inheritance of acquired characters, he had rather mild expectations for the ability of his own evidence to change minds, even among his close friends: he writes to Hooker, for instance, that "[w]henever my book on poultry, pigeons, ducks, and rabbits is published, with all the measurements and weighings of bones, I think you will see that 'use and disuse' have at least some effect" ([March] 26 [1862] in MLD v. 1 199). Thus, Darwin could not have failed to recognize that a theory of generation and inheritance would not have needed to allow for the inheritance of acquired characters to count as a serious contender even in his own day.

What emerges from this lengthy discussion is that Darwin's acceptance of the inheritance of acquired characters certainly posed no insurmountable obstacle and perhaps not even a serious one to recognizing or accepting the possibility of a common-cause structure for inheritance. Such an alternative could have preserved most of the explanatory accomplishments of pangenesis itself, even bettering them in some cases, and the cases of the inheritance of acquired characters Darwin found convincing could either have been reconciled with a common cause structure for inheritance in a manner analogous to that used for distant reversion or simply left as empirical anomalies for the theory, as he was happy to do with other troubling phenomena equally or even more widely accepted by the scientific community of his time. Thus, when Darwin repeatedly insists that pangenesis is the only account he knows that can explain the phenomena of generation and heredity, we should take him at his word and conclude that he failed to conceive of even the possibility of any common-cause alternative to pangenesis in the first place.

3.3 Darwin's Failure to Grasp Galton's Common Cause Mechanism for Inheritance

By this point it will surely seem to some readers that I have already spilled an undue amount of ink defending the rather modest historical thesis that

Darwin never conceived of the possibility of a common-cause structure for inheritance or a common-cause mechanism of hereditary resemblance, but even this unassuming claim must still face at least one daunting historiographical challenge: how are we to reconcile it with the fact that the earliest expressions of the theory of the continuity of the germ plasm reach back perhaps as far as Richard Owen's 1849 work on parthenogenesis (see Thompson 1888–1889, cited in Robinson 1979 30n) and in any case certainly to Francis Galton's 1865 article "Hereditary Talent and Character" in *Macmillan's* magazine? There is little room to doubt that Darwin read Galton's article, for he refers readers of VAP (e.g. the American edition of 1868) to this "able paper on hereditary talent" (v. 2 16). And the claim in Galton's paper most noted by historians of science is the following startling suggestion:

> We shall therefore take an approximately correct view of the origin of our life, if we consider our own embryos to have sprung immediately from those embryos whence our parents were developed, and these from the embryos of *their* parents, and so on forever. We should in this way look on the nature of mankind, and perhaps on that of the whole animated creation, as one continuous system, ever pushing out new branches in all directions, that variously interlace, and that bud into separate lives at every point of interlacement. (Galton 1865 322)

We should not, however, make the mistake of assuming simply because Darwin read Galton's 1865 paper that he either recognized or understood the idea of the continuity of the germ plasm proposed therein. The central aim of the 1865 article was not to propose a mechanism or theory of inheritance at all, but instead to establish the non-inheritance of acquired mental abilities in human beings and (to borrow Cowan's appealing term) the 'omnicompetence' of heredity in determining human mental and moral characteristics.[18] Perhaps unsurprisingly, then, the use Darwin makes of this paper is only to suggest that while some "have doubted whether those complex mental attributes, on which genius and talent depend, are inherited.... he who will read Mr. Galton's able paper on hereditary talent will have his doubts allayed" (VAP (first American edition, 1868) v. 2 16; the second edition of VAP (v. 1 538) mentions instead, in an otherwise identical passage, "Mr. Galton's able work on 'Hereditary Genius'," a reference to Galton's 1869 book of that name). This does not yet, of course, provide any evidence that Darwin failed to understand Galton's idea of the continuity of the germ plasm, but it does show why Darwin's mention and apparently favorable opinion of the 1865 article need not be taken as evidence of having considered or understood the paper's brief and tangential suggestion of germ line continuity.

Furthermore, there is indeed telling evidence of Darwin's failure to comprehend Galton's proposal of the continuity of the germ plasm in their exchange of correspondence of 1875, preceding Galton's presentation of his paper "A Theory of Heredity" to the Anthropological Institute. Hearing of Galton's interest in the matter and impending publication, Darwin wrote in early November of 1875 to warn him of Huxley's distrust of the views of Balbiani (all the correspondence in this exchange can be found in Pearson's *Life Letters and Labours of Francis Galton* (hereafter LLL) 1914–1930 v. 2 181–189). Galton's appreciative reply sought to summarize the contents of "A Theory of Heredity," including his view that:

> we must not look upon those germs that achieve development as the main sources of fertility; on the contrary, considering the far greater number of germs in the latent state, the influence of the former, i.e. of the personal structure, is relatively insignificant. Nay further, it is comparatively sterile, as the germ once fairly developed is passive; while that which remains latent continues to multiply. (LLL 182)

By elaborating this view, Galton claims to account "both for the fact, and for the great rarity and slowness of the inheritance of acquired modifications" (LLL 182). He then suggests that the appropriate analogy for the relationship between parent and child is not that of parent country to colonists, but of the representative government of the parent nation to that of the colonists, under the supposition that a small proportion of the colonists are nominated to its government by the government of the parent country. With this, Galton says, he has "so far as the limits of a letter admit, made a clean breast of my audacity in theoretically differing from Pangenesis," a difference he summarizes with the following two propositions:

1. In supposing the sexual elements to be of as early an origin as any part of the body (it was the emphatic declarations of Balbiani on this point that chiefly attracted my interest) and that they are not formed by aggregation of germs, floating loose and freely circulating in the system, and
2. In supposing the personal structure to be of very secondary importance in Heredity, being, as I take it, a *sample* of that which is of primary importance, but not the thing itself. (LLL 183, original emphasis)

Although Darwin was "delighted that you stick up for germs," he seems to have been unable to follow Galton's summary, saying only that he "can hardly form any opinion until I read your paper *in extenso*" (and drawing Galton's attention to Brown-Séquard's latest experimental results supporting the inheritance of acquired characters and to "the many cases of parthenogenesis"). He reports that he is "very glad indeed of your work, though I cannot yet follow all your reasoning" (Darwin to Galton, Nov. 4, 1875; in LLL 183). Galton responded to this invitation by sending Darwin one of the proofs of "A Theory of Heredity" with the "hope it will make my meaning more clear." There again Galton had proposed the continuity of the germ plasm, defining an organism's 'stirp' as "the sum total of the germs, gemmules, or whatever they may be called, which are to be found, according to every theory of organic units, in the newly fertilized ovum—that is in the early pre-embryonic stage—from which time it receives nothing further from its parents, not even from its mother, than mere nutriment" (LLL 185), and he argued that "The stirp of the child may be considered to have descended directly from a part of the stirps of each of its parents, but then the personal structure of the child is no more than an imperfect representation of his own stirp, and the personal structure of each of the parents is no more than an imperfect representation of each of their own stirps" (LLL 186).

Darwin found the paper itself no easier to grasp and no less puzzling than Galton's summary had been. He writes:

> I have read your essay with much curiosity and interest, but you probably have no idea how excessively difficult it is to understand. I cannot fully grasp, only here and there conjecture, what are the points on which we differ—I daresay this is chiefly due to muddle-headiness[19] on my part, but I do not think wholly so. Your many terms, not defined "developed germs"—"fertile" and "sterile" germs (the word 'germ' itself from association misleading to me), "stirp,"—"sept," "residue" etc. etc., quite confounded me.... Unless you can make several parts clearer, I believe (though I hope I am altogether wrong) that very few will endeavor or succeed in fathoming your meaning. (LLL 187)

What followed this letter of November 7 was an exchange in which Darwin tried to explain the sources of his confusion and skepticism while Galton sought unsuccessfully to make his position clear to Darwin (see Cowan 1985 117–118). It appears from the letters that part of the dispute was mediated by George Howard Darwin, traveling between London and Down representing the views of each correspondent to the other in person (see Darwin's letter of Dec. 18 and Cowan 1985 118). At no point in this exchange did Darwin show any evidence of having resolved his initial perplexity or of understanding the idea of the continuity of the germ plasm that Galton sought to propose, and although he appreciated the gracious spirit in which Galton had received his earlier accusations of obscurity, he nonetheless persisted in finding his cousin's account of heredity opaque:

> I have this minute finished your article in Fraser[20] and I do not think I have read anything more curious in my life.... I should be glad to be convinced that the obscurity was *all* in my head, but I cannot think so, for a clear-headed (clearer than I am) member of my family read the article and was as much puzzled as I was. To this minute I cannot define what are "developed," "sterile" and "fertile" germs. You are a real Christian if you do not hate me for ever and ever." (Darwin to Galton, Nov. 10, 1875; in LLL 188–189)[21]

The irony in this situation, of course, is that Darwin was anything but alone in failing to grasp the fundamental structure of inheritance that Galton sought to propose. After all, the doctrine of the continuity of the germ plasm is most famously associated not with Galton but with August Weismann: although Weismann acknowledged that Galton had recognized the possibility of the continuity of the germ plasm, he was surely right to suggest that this idea had enjoyed virtually no attention and was of little significance for the scientific community at the time (Robinson 1979 30; cf. Dunn 1965 38–39). The doctrine of the continuity of the germ plasm and the correlative idea that phenotypic continuities between ancestors and offspring might be the results of a common cause rather than links in a causal chain was simply not a feature of the scientific landscape prior to Weismann's publication of it in 1883.

Of course, the central issue before us is not whether Darwin would have accepted Galton's proposed continuity of the germ plasm had he understood it—he almost certainly would not have, in large part because he was increasingly convinced of the widespread existence and importance of the inheritance of acquired characters, whose significance Galton sought to minimize. Furthermore, Darwin was looking for a theory of inheritance that would permit natural selection to function as the engine of evolution, and he took this to require allowing an important role for the inheritance of acquired characters (though see note 15 above).[22] The point is instead that Darwin shows no evidence of having considered and rejected the idea that similarities between ancestors and offspring might be results of a common cause rather than links in a causal chain, and indeed shows no evidence of even having been able to understand this line of thought when it was explicitly presented to him directly by Galton. Instead the most natural conclusion to draw from the historical evidence is that Darwin simply failed to conceive of or consider the entire *class* of theoretical alternatives to pangenesis picked out by this idea, notwithstanding the fact that it offered an equally promising strategy for explaining what he took to be the central phenomena of inheritance and generation.[23] And as we will see in the next two chapters, it is not only our own theories of genetics and embryology that must be counted among this broad class of theoretical alternatives unconceived and unconsidered by Darwin, but also others, like those of Galton and Weismann, involving (by present lights) fundamental errors about the nature of inheritance and generation (as well as failures to conceive of serious alternative possibilities) no less profound than those implicated in pangenesis itself.

Notes

1. Here and throughout I have tried to restrict my use of the secondary literature concerning this period in the history of science to classic discussions in

the field whose central contentions still appear to be widely accepted, rather than to the unavoidably more contentious claims embodied in more recent historical scholarship. As will become clear in what follows, however, the direct evidence I adduce in support of the problem of unconceived alternatives in the work of Darwin, Galton, and Weismann is drawn almost exclusively from primary sources, rather than from this secondary literature. Of course, if more recent developments in the historical scholarship concerning this period undermine either my reading of the primary sources or the use I have made of them in trying to establish the general significance of the problem of unconceived alternatives itself, I trust that my colleagues in the history of science will set me straight.

2. As Mazzolini and Roe point out (1986 21–22), Spallanzani's chief epigenesist opponent John Turberville Needham did not defend true spontaneous or "equivocal" generation, equated in his own day with the formation of organisms by accidental or chance events. Instead, Needham argued that the formation of tiny animalcules from nonliving material took place in an orderly and regular succession according to divine laws.

3. While even preformationists allowed a role for natural forces in the processes of generation (i.e., in initiating the growth of an embryo), they denied that such forces could operate in a *building* or *formative* fashion (see Mazzolini and Roe 1986 31–32, Roe 1981 chap. 2).

4. A terminological caution: the term 'particulate' heredity is often used to describe views on which specific *characteristics* or the material foundations for them are inherited in a discrete fashion, that is, in opposition to 'blending' heredity (in which parental characteristics or their material causes are mixed or amalgamated in the offspring). While views like Darwin's and Spencer's certainly involved the postulation of material particles inherited by offspring from parents, they were not particulate views of heredity in this important sense. Indeed, the blending of parental *traits* in inheritance was a prominent feature of Darwin's and of many later nineteenth-century accounts of generation that conceived of material particles passed from parents to offspring as the foundation of heredity, and as Keith Benson has stressed to me in private correspondence, blending notions of inheritance in one form or another would play an important role in the work of all three major figures we will consider. Cf. also Michael Bulmer's extremely useful distinction between "phenotypic" and "physical" blending (2004 293).

5. Except where otherwise noted, page numbers will refer to the 1905 republication of the second edition of this work as a "popular edition" by the original publisher, John Murray.

6. In August of 1867 Darwin wrote to Charles Lyell "I do not know whether you have ever had the feeling of having thought so much over a subject that you had lost all power of judging it. This is my case with Pangenesis (which is 26 or 27 years old), but I am inclined to think that if it be admitted as a probable hypothesis it will be a somewhat important step in Biology" (Darwin to Lyell, August 22 [1867] in *Life and Letters of Charles Darwin* (1959 [1887]; hereafter LLD), v. 2 255). Further compelling evidence that Darwin was a "lifelong generation theorist" is provided by Hodge (1985; discussed in Bowler (1989 58); see also Endersby (2003); cf. Geison (1969)). This evidence weighs equally against the "Franciscan" view (so named for Darwin's son Francis, who put it forward in LLD) that pangenesis was a late addition to the main currents of Darwin's evolutionary thought.

7. To Hooker Darwin wrote: "H. Spencer says the view is quite different from his (and this is a great relief to me, as I feared to be accused of plagiarism, but utterly failed to be sure what he meant, so thought it safest to give my view as almost the same as his), and he says he is not sure he understands it..." (Darwin to Hooker, February 23 [1868] in LLD v. 2 259–261). To Wallace he wrote: "I now hear from H. Spencer that his views quoted in my foot-note refer to something quite distinct, as you seem to have perceived" (Darwin to Wallace, February 27 [1868] in LLD v. 2 262). As a general matter, the similarities between Darwin's account and any number of earlier views of inheritance appealing to material units or particles seem to have caught him somewhat by surprise (see letters from Darwin to Huxley, July 12 [1865?] in LLD v. 2 228, Darwin to Huxley, [1865?] in LLD v. 2 228–229, and Darwin to Ogle, March 6 [1868] in LLD v. 2 265): the pangenesis manuscript of 1865 contained no mention of related earlier theories, the first edition of VAP discussed those of Buffon, Bonnet, Spencer, and Owen, and the second edition added mention of more views "nearly similar" to pangenesis by Hippocrates, Ray, and a Prof. Mantegazza (VAP v. 2 457n; see Geison 1969 393). Of course, the fact that the *general* suggestion of hereditary particles thrown off by parts of the body had been previously made should not lead us to think that pangenesis itself was not really new or was not genuinely unconceived before Darwin's work in the mid-nineteenth century: Darwin himself was quick to point out important differences between pangenesis and these earlier conceptions (see, e.g., VAP v. 2 457n and Darwin to Huxley, [1865?] in LLD v. 2 228–229)—with the curious exception of that of Hippocrates (Darwin to Ogle, March 6 [1868] in LLD v. 2 265)—and as Geison notes, "Darwin could probably have demonstrated... fundamental differences between his ideas and those of any of the pre-nineteenth century pangenetic theorists" (1969 395).

8. One consequence of this view is that the central problem of inheritance for Darwin (and for Galton as well) was to account for the source of new variation. Indeed, some later alternatives we will consider may have remained unconceived by Darwin precisely because he saw any successful theory of generation as needing to account for both inheritance and variation simultaneously.

9. From a letter written to Darwin himself (February 1868 in MLD v. 1 300).

10. And as late as 1873 Darwin confessed to A. De Candolle that "[a]lthough my hypothesis of pangenesis has been reviled on all sides, yet I must still look at generation under this point of view..." (Darwin to De Candolle, January 18 [1873] in MLD v. 1 348).

11. I defer for the moment discussing the possibility that given the phenomena he took to exist, Darwin was *right* to think that (some version of) pangenesis alone could offer a convincing explanation for them.

12. My thanks to the Cambridge University Library for providing me a reproduction of Darwin's unpublished letter to Hooker, and to the Staatsbibliothek zu Berlin—Preußischer Kulturbesitz for providing me with a reproduction of Darwin's unpublished letter to Carus (Slg. Darmst. Lc 1859 (9): Darwin, Charles Robert—Brief vom 19.10 [1868] an Victor Carus [=Br. Nr. 14]).

13. This reservation is somewhat surprising, for in another letter to Hooker just five days later Darwin would write that pangenesis' singular explanatory achievements lead him to "fully believe that each cell does *actually* throw off an

atom or gemmule of its contents" (cf. Darwin's letter to Hooker of Feb. 28 [1868] in LLD v.2 264, quoted and discussed above).

14. The last two phenomena actually provide a nice illustration of one specific way in which the original pessimistic induction's willingness to project from past to present science is too simple, for much of the evidence of these phenomena for which Darwin was concerned to account was gathered from famous anecdotes (such as that of Lord Morton's chestnut mare; see VAP ii 446, MLD 2 359), folk wisdom, the stories of animal breeders, and the like (see Olby 1985 44 and 79, where the mare's owner is given as Lord Moreton), while the concerted efforts of more recent scientific methodology have undoubtedly established more stringent standards for the collection of data. But this difference does not mitigate the problem of unconceived alternatives, as Darwin was unable to exhaust the space of convincing potential explanations (by his own standards) of the phenomena for which *he* thought a theory of generation needed to account.

15. A note of caution is in order here, however. Winther notes (2000 436–439) that Darwin felt increasingly forced to make room in his theory for a source of systematic, directed, nonrandom, or necessarily adaptive variation (including the inheritance of acquired characteristics) by the need to publicly accept Kelvin's estimate of the age of the Earth (which seemed to allow insufficient time for natural selection to produce present organismic diversity from a pool of purely random variation; see also Gayon 1998 87–88) which he privately rejected. Furthermore, Darwin clearly saw the danger thus posed to the theory of natural selection: as he wrote to Asa Gray in 1868, "If the right variations occurred, and no others, natural selection would be superfluous" (cited in Winther 2000 439; see also Gayon 1998 54). Thus, while Darwin was certainly convinced (along with many other naturalists of the nineteenth century) that the inheritance of acquired characters occurred, it would be easy to overestimate the importance he sincerely ascribed to this mechanism on the basis of the second edition of the *Variation* and other late published writings.

16. Darwin seems to recognize this, concluding merely that we have gained "some insight" (VAP v. 2 488) into distant reversion and ultimately that "[r]eversion depends on the transmission from the forefather to his descendants of dormant gemmules, which occasionally become developed under certain known or unknown conditions" (VAP v. 2 491). In the pangenesis manuscript of 1865 he simply attributes distant reversion to "unknown causes" (Olby 1963 261), and to Hooker he writes that "crossing races as well as species tends to bring back characters which existed in progenitors hundreds and even thousands of years ago. Why this should be so, God knows" ([September 13, 1864] in MLD v. 2 339–340). Nonetheless, the seriousness with which Darwin regarded the demand to explain distant reversion is well illustrated by his reaction to Naudin's account of hybrids as 'living mosaics' without any true fusion of elements from the crossed species: in the margin of his copy of Naudin's prizewinning 1862 essay on hybrids he writes simply "This view will not account for distant reversion" (Olby 1985 51) and to Hooker he writes that he "cannot think that [Naudin's view] will hold" giving as his only reason that it "throws no light, that I can see, on this reversion of long-lost characters" ([September 13, 1864] in MLD v. 2 339–340). Nonetheless, Darwin does explicitly follow Naudin's account of reversion in the offspring of ordinary hybrids in VAP v. 2 486–487. On the importance of reversion for Darwin, see also Gayon 1998 44–45.

17. Cf. also Olby (1985 58 original emphasis): "as Cowan admits . . . there was little *hard* evidence at that time in support of the inheritance of acquired characters."

18. Galton seeks to discharge this task in a remarkably dogmatic way, offering little in the way of scientific argument or evidence and much in the way of generalities, assurances, and eugenic fantasies. Indeed, Cowan (1985 65–66) describes the 1865 article as "a failure as a scientific treatise" and "an exercise in political rhetoric," taking it to illustrate that both Galton's interest in heredity and his commitments on contentious matters of fact were rooted in his eugenic ambitions.

19. MLD v. 1 360 has "muddy-headedness."

20. *Fraser's Magazine*, in which Galton had published articles concerning heredity in 1873 and 1875.

21. Alison Pearn of the Darwin Correspondence Project has suggested to me that this "clear-headed" member of Darwin's family was probably his daughter Henrietta, to whom he often showed materials he wished to discuss. Notice that in both this and Darwin's earlier letter to Galton, the specific terms Darwin singles out as central to his confusion are the ones at the heart of the common cause character of the stirp theory itself: "fertile" and "sterile" germs, "developed" germs, "stirp" and "residue". Thus it does not seem plausible to suggest that in fact Darwin grasped Galton's fundamental idea of a common cause account of inheritance and that it was instead some other aspect of the stirp theory he found so perplexing.

22. Note that while the class of common-cause alternatives neglected by Darwin certainly includes some members (like Mendelian or contemporary genetics) with particulate (in the sense of nonblending) heredity, we should not make the mistake of trying to support the problem of unconceived alternatives by appealing to the widespread presumption that a particulate conception of heredity was somehow the natural complement to Darwin's selectionist conception of evolution or the missing piece of a seamless puzzle and suggesting that Darwin would surely have embraced particulate heredity as the bride of natural selection if only he had thought of it. As Bowler argues convincingly (1989 61–63), this suggestion depends upon a misreading of Darwin's response to Fleeming Jenkin's famous argument that blending inheritance makes evolution impossible (because characteristics that arose and were favored by selection would be swamped by blending in subsequent matings) and a misunderstanding of Darwin's commitment to both the gradual character of the process of evolution and the continuous (rather than saltational) character of the traits on which selection acts. (For useful detailed discussion, see Gayon 1998 chap. 3, and Bulmer 2004.) The integrity of Darwin's biological theorizing as a whole, and the consequent implausibility of seeing his evolutionary thought as naturally completed only by later genetic theories, is also convincingly defended by Hodge (1985 esp. 241–243).

23. Of course, the class of common cause accounts of inheritance was not the *only* set of serious alternatives that escaped Darwin's notice, and the broader point here is not Darwin's failure to consider this specific set of hypotheses but rather his failure or inability to exhaust the space of serious alternative possibilities *generally*. Still, the importance and centrality of this particular set of possibilities both to one of Darwin's own contemporaries (Galton) and to immediately subsequent theorizing about inheritance and generation make it impossible to argue that it wasn't really a serious competitor by the standards of the time.

4

Galton and the Stirp Theory

[A]ll thought processes and thought-constructs appear a priori to be not essentially rationalistic, but biological phenomena.... Thought is originally only a means in the struggle for existence and to this extent a biological function.

—Hans Vaihinger, *The Philosophy of 'As if'*

4.1 The Transfusion Experiments: "A Dreadful Disappointment to Them Both"

Galton would fully develop the alternative physiological conception of heredity prefigured in "Hereditary Talent and Character" only after relaxing his opposition to the inheritance of acquired characters long enough to conduct a now-famous series of transfusion experiments in hopes of vindicating pangenesis itself. It was only after these experiments spectacularly failed to confirm pangenesis that Galton abandoned Darwin's hypothesis altogether, retaining its mechanism only in the explicitly subsidiary role of explaining any cases of the inheritance of acquired characters he might ultimately be forced to accept and turning his primary attention to developing what would become his own "stirp" theory of inheritance instead.

In these transfusion experiments,[1] Galton introduced into purebred "silver grey" rabbits blood from various other breeds. He reasoned that if pangenesis were true, the gemmules passed on to the progeny of silver greys thus transfused would be a mix of those derived from their parents' tissues and those added from the blood of the other breeds, and that any such progeny should therefore exhibit some "mongrelism": variations in traits or features that were uncharacteristic of the pure breed. Galton hoped to vindicate pangenesis by observing such changes, and he saw in this the promise of a general procedure for manipulating the characteristics of animals:

> If Pangenesis were true, according to the interpretation which I have put on it, the results would be startling in their novelty, and of no small practical use; for it would become possible to modify varieties of animals, by introducing slight dashes of new blood, in many ways important to breeders. Thus, supposing a small infusion of bull-dog blood was wanted in a breed of greyhounds, this or any more complicated admixture, might be effected... in a single generation. (1870–1871 395)

These hopes would be disappointed, however, as the offspring of the transfused rabbits stubbornly persisted in displaying only the characteristic features of the breed along with quite typical variations (on which Galton sometimes seized with great enthusiasm before learning that they were commonly found in silver greys; see, e.g., Pearson's *Life Letters and Labours of Francis Galton* (hereafter LLL) 1914–1930 II 159). Galton responded by improving his experimental techniques and replacing ever-higher proportions of the blood of the silver greys[2] but with no success. In a paper presented to the Royal Society (1870–1871), he reluctantly concluded that pangenesis had failed his test.

There is no question that Darwin was well aware of the presumptions under which Galton was proceeding in the transfusion experiments, most importantly, the assumption that blood was the vehicle of circulation for gemmules in the bodies of animals. Not only was this presumption clearly stated in *Hereditary Genius* (which Darwin read and referred to in the second edition of *The Variation of Animals and Plants Under Domestication* (hereafter VAP), see chap. 3), but the two men engaged in an extended correspondence about the course of the transfusion experiments from their first conception, when Galton wrote Darwin in December of 1869 to ask for his help in trying to obtain purebred rabbits. Of their correspondence concerning the experiments from this time to the publication of Galton's negative results to the Royal Society in 1871, only Galton's side survives, but it is clear from these letters that Darwin took an active interest in the course of the transfusion experiments and that he regularly advised and encouraged Galton throughout the course of this research in hopes of finding evidence to support pangenesis (see LLL II 156–169). Furthermore, a letter from Mrs. Darwin to her daughter Henrietta in March of 1870 reports that "F. Galton's experiments about rabbits (viz. injecting black rabbit's blood into grey and *vice versa*) are failing, which is a dreadful disappointment to them both" (LLL II 158).

Following Galton's publication of his results to the Royal Society, however, Darwin published a letter in *Nature* refusing to grant conclusive force to the negative outcome of Galton's transfusion experiments, insisting that:

> in the chapter on Pangenesis... I have not said one word about the blood, or about any fluid proper to any circulating system. It is, indeed, obvious that the presence of gemmules in the blood can form no

> necessary part of my hypothesis; for I refer in illustration of it to the lowest animals, such as the Protozoa, which do not possess blood or any vessels; and I refer to plants in which the fluid, when present in the vessels, cannot be considered as true blood. (*Nature* April 27, 1871; reprinted in LLL II 163–164)

Historians of science have not been kind to Darwin in this connection. Cowan says that Darwin "behaved badly" in thus responding to the transfusion experiments after having "cooperated fully" in them and having "never hinted at the possibility that the experiments might not be conclusive" (1985 112–113). Robinson mutters darkly that the episode shows "one does not always know about Darwin" (1979 9), and Pearson seems to regard Galton's forbearance in responding to this betrayal as one of the most admirable acts of his life (LLL II 157, 165). In reply to Darwin, Galton published a remarkably deferential and genial letter of his own in *Nature* (reprinted in LLL II 164–165) in which he accepted Darwin's claim "that the views contradicted by my experiments . . . differ from those [Darwin] entertained." Likewise in a private letter to Darwin (April 25, 1871, in LLL II 162) he apologized for having misunderstood the theory rather than complaining of having been mislead by Darwin's involvement and enthusiasm into thinking the transfusion experiments offered a suitable test of pangenesis. In both letters he did go on to point out passages in VAP in which Darwin's use of terms like "circulate," "freely," and "diffusion" had suggested to him that blood was supposed to be the circulating medium for gemmules (and in the published letter he offered suggestions for their reformulation[3]), but he never suggested that Darwin was now changing his mind; the situation, he claimed was more:

> [a]s if, having heard my trusted leader utter a cry, not particularly well articulated, but to my ears more like that of a hyena than any other animal, and seeing none of my companions stir a step, I had, like a loyal member of the flock, dashed down a path of which I had happily caught sight, into the plain below, followed by the approving nods and kindly grunts of my wise and most respected chief. And I now feel, after returning from my hard expedition, full of information that the suspected danger was a mistake, for there was no sign of a hyena anywhere in the neighborhood. I am given to understand for the first time that my leader's cry had no reference to a hyena down in the plain, but to a leopard somewhere up in the trees; his throat had been a little out of order—that was all. Well, my labour has not been in vain; it is something to have established the fact that there are no hyenas in the plain, and I think I see my way to a good position for a look out for leopards among the branches of the trees. In the meantime, Vive Pangenesis! (*Nature* May 4, 1871, reprinted in LLL II 164–165)

Galton left it to Darwin to consider just how frequent and forceful his "approving nods and kindly grunts" had been. And perhaps most surprisingly of all, he continued to involve Darwin in the transfusion experiments as they continued (with uniformly negative results; see VAP (2d edition), 350n (cited in LLL II 177) and Galton's "A Theory of Heredity," 342n) for at least the next three years, even arranging for Darwin to house, care for, and mate the experimental rabbits during Galton's travels![4]

The gentle tone and judicious character of Galton's response is undoubtedly impressive, especially given Darwin's willingness to claim publicly that Galton had misunderstood pangenesis without (apparently) ever having suggested this in private and after having been at the very least an enthusiastic advisor to the experiments themselves. But it is not clear that Darwin deserves the abuse historians have heaped upon him for responding to Galton's transfusion experiments as he did. For one thing, Darwin's letter to *Nature* concedes that he himself shared in the mistake he attributes to Galton, saying "when I first heard of Mr. Galton's experiments, I did not sufficiently reflect on the subject, and saw not the difficulty of believing in the presence of gemmules in the blood" and he also explicitly recognizes himself as having missed or lost track of his original meaning for a time, saying "when I used these latter words ['circulation of fluids throughout the body'] and other similar ones, I presume that I was thinking of the diffusion of the gemmules through the tissues or from cell to cell, independently of the presence of vessels" (*Nature* April 27, 1871; reprinted in LLL II 163–164). In this context, Darwin's claim that it is "obvious that the presence of gemmules in the blood can form no necessary part of my hypothesis" can hardly be taken as intended to suggest that this was *always* obvious either to himself or to everyone but Galton. Instead he seems most concerned to point out simply that the hypothesis of pangenesis *itself* need not assume that gemmules are circulated in the blood, whether he or Galton had noticed this or not, and that the theory itself is therefore not definitively refuted by the transfusion experiments.[5] And the terms of his refusal to grant dispositive force to these experiments are commensurately mild: Darwin allows that Galton's experiments are "extremely curious, and that he deserves the highest credit for his ingenuity and perseverance;" insisting only that it is "a little hasty" for Galton to infer from them the falsity of pangenesis "beyond all doubt" (Galton's words (1870–1871), quoted in Darwin's letter) and that "it does not appear to me that Pangenesis has, as yet, received its death blow." And Darwin concludes by offering the vulnerability of pangenesis as an "excuse for having said a few words in its defence." Of course, noting these aspects of Darwin's response may help to explain why Galton was neither moved to a more indignant reaction nor even disinclined to involve Darwin in further pursuing the transfusion experiments themselves.

Whatever our judgment of the character of Darwin's conduct on this occasion, the episode itself suggests that the new induction and the problem of unconceived alternatives may also be central to any real argumentative force we should ultimately concede to the *holist's* distinctive challenge to scientific realism. That challenge, recall, began from the homely observation that scientific theories about inaccessible domains of nature cannot be tested in isolation but instead must be conjoined with auxiliary hypotheses in order to generate testable predictions or other observable consequences. But from this the holist draws the dramatic conclusion that the failure of any given experimental prediction invariably leaves open a choice between rejecting the theory we set out to test or one of the auxiliary hypotheses we used to test it instead. The holist insists, therefore, at least that no amount of evidence nor any particular piece of evidence can ever rationally compel us to reject a given scientific theory, and perhaps even that belief in any given theory can be rationally maintained in light of any evidential findings whatsoever ("come what may"). In this case, Galton's efforts to test pangenesis relied on auxiliary assumptions explicitly including the claim that the gemmules of animals would circulate in their blood, and Darwin pointed out that we could reconcile pangenesis with the negative outcome of the transfusion experiments by blaming this auxiliary hypothesis for the negative experimental results rather than giving up pangenesis itself.[6]

What is suggestive about this historical sequence of events, however, is that Darwin seems never to have seriously contemplated the possibility that gemmules might be thrown off by the developed tissues of adult animals but not circulated in their blood until he was *forced* to do so by the negative outcome of Galton's transfusion experiments. Only under the threat of such empirical refutation was Darwin even motivated to conceive of the alternative suggestion that gemmules might instead be diffused "through the tissues or from cell to cell, independently of the presence of vessels" (1871; reprinted in LLL II 163–164). Or, if we are to believe the interpretation Darwin offers of his earlier self, it was at just this point that he was prompted to rediscover what he originally had in mind when he proposed the theory. But in either case, this possibility was not among the alternatives that either Galton or Darwin had in mind when the transfusion experiments were actually being conducted.

The idea here is certainly not that unconceived alternatives generally are well-represented by the (disturbingly vague) alternative Darwin proposes, still less that disagreements about whether a particular theory should be abandoned in light of particular evidential findings *must* involve alternative possibilities unconceived in advance of the findings in question. Nonetheless, this case illustrates how a version of the holist's traditional worry can be grounded in the general problem of unconceived alternatives and offers at least one genuine historical precedent. Perhaps most importantly, only this version of the problem threatens to put any real teeth into the holist's concern that it might be rationally defensible to maintain belief in a given theory in light of absolutely any particular evidential findings or even hold onto a theory "come what may," because it suggests that there may well be *plausible* modifications of a theory we are testing or *serious* alternative auxiliary hypotheses available that would reconcile it with the available evidence *even if we cannot presently conceive of what these might be.* After all, if we had good reason to think that we were able to consider all the serious or plausible versions of any given theoretical proposal (or all the serious auxiliary hypotheses with which it might be conjoined to generate testable predictions), then the holist's predicament would always be at least *tractable*: to refute a given theoretical idea we would simply need to eliminate all the serious alternative formulations of the idea or test all the combinations of the hypothesis with serious alternative auxiliary hypotheses.[7] But if instead we have good evidence that scientifically serious modifications of a theory (or scientifically serious alternative auxiliary hypotheses) routinely remain unconceived by us—as the case of Darwin's response to the transfusion experiments might suggest—then it really might be reasonable (or at least rationally unimpeachable) to hold on to our belief in a successful theory's central contentions in light of evidential disconfirmation *even if we cannot presently formulate the serious modifications or alternative auxiliary hypotheses that would enable these central contentions to escape refutation.* And in this case the holist will be in a position to claim that no one response to experimental disconfirmation can ever be singled out as uniquely rational, even for the reasonable (rather than radically skeptical) enquirer. To put the matter another way, if we believe that there are an unknown, indefinite, or indeterminate number of presently unconceived serious ways of modifying a given hypothesis in light of recalcitrant experimental data, this would seem to leave us unable to effectively assess the vulnerability of a given hypothesis to particular evidential findings.

On the other hand, the historical facts of this particular case offer little encouragement to those thinkers who insist (perhaps most influentially in the contemporary Sociology of Scientific Knowledge movement) that because the response to empirical refutation or disconfirmation is underdetermined in the way the holist suggests, the viability of such a challenge to or defense of a given theory is invariably a matter of social negotiation whose outcome depends on the comparative power or resources of the competing groups of social actors who find it in their interests to pursue it. Not even Darwin's famous network of powerful scientific allies seem to have regarded his suggestion in response to the transfusion experiments that gemmules might be diffused independently of the blood vessels as the least bit convincing, and even Galton (whose personal loyalties, not to mention his clearest route to authority, advancement, and social power in the context of nineteenth-century British science, clearly lay with Darwin himself[8]) went on to consider instead the suggestion that the gemmules of pangenesis might be only *temporary* residents of the blood (see LLL II 161). While William Keith Brooks would later attempt to revive pangenesis, he would simply preserve the assumption that blood was the circulating medium for gemmules (see Robinson 1979 chap. 5). And indeed, even when Galton's development of the stirp theory ultimately did lead him to accept the possibility that germs could sometimes "transgress the bounds of the cell or cell-interspace in which their progenitors had lodged" just as "a blood-corpuscule will occasionally find its way through the unruptured wall of a capillary vessel," he made a point of insisting that this was "a very different supposition to that of the free circulation of gemmules in Pangenesis" and continued to argue that "[o]n physical grounds, we cannot understand how colloid bodies, such as the Pangenetic gemmules must be, could pass freely through membranes" (1876 341). It is hard to imagine either a figure enjoying greater authority, power, or social resources in late nineteenth-century biological science than Darwin, or a less successful effort to blunt the experimental refutation of a theory by challenging the auxiliary hypotheses on which it depended.[9]

4.2 Galton's Stirp Theory and Its Maturational, Invariant Conception of Heredity

In any case, following what he seems to have regarded as a decisive refutation of pangenesis by his transfusion experiments, Galton returned to the fundamental ideas about heredity originally bruited in "Hereditary Talent and Character." In a series of publications including, most importantly, the articles "On Blood Relationship" (hereafter BR) and "A Theory of Heredity" (TH),[10] Galton's recognition of the possibility of a common cause structure for inheritance blossomed into his own "stirp" (from the Latin '*stirpes*,' a root) theory of the mechanisms of inheritance and generation. Darwinian pangenesis had sought to explain reversion (including the nonappearance of female secondary sexual characteristics in males and vice versa) by appealing to the somewhat vague idea that gemmules could sometimes remain "latent" in an organism and thus be transmitted from an ancestor to a later descendant without being expressed in the intervening generation(s). Galton seized upon this idea of latency and extended it into a general common-cause mechanism of inheritance by proposing that a body's germinal elements were segregated into two streams. Immediately after fertilization, he argued, the entire "stirp" of inherited germinal materials ("the sum-total of the germs, gemmules, or whatever they may be called, which are to be found, according to every theory of organic units, in the newly fertilized ovum," TH 330) is separated by a process of "Class Representation" into 'patent' elements that will develop into the various constituent parts of the adult organism and a much more numerous residue of 'latent' elements that will instead remain undeveloped.[11] After the constituents of this residue have been multiplied, a process of "Family Representation" acts to segregate out those latent elements that will be passed along to the organism's offspring during reproduction from the latent elements that will die with the individual.[12] Once the former group is united with the corresponding latent elements from another parent (forming a stirp of equal size to that with which the process began in each parent) the resulting stirp is again separated in turn into patent gemmules from which the constituents of the offspring's own body will develop and the stock of latent gemmules from which the stirp it passes along to the next generation will be selected after further multiplication, and so on. Galton thus sought to explain reversion, individual variation, and the resemblances of offspring to their parents and siblings as statistical effects of a single physiological process of heredity. He preserved the bare possibility that patent gemmules might also occasionally contribute progeny to the stirp (and acquired characteristics of the parents thereby be reproduced in offspring in the way pangenesis allowed), but relegated this process to an "unimportant" (BR 400) and "supplementary and subordinate" (TH 330) role[13] with "minute and secondary" (TH 339) effects on inheritance, explicitly intended only to account for whatever evidence of the inheritance of acquired characters could not be otherwise explained or explained away (see BR 398–400, TH passim, esp. 342).

It is worth noting that Galton does not propose this common-cause mechanism for heredity in a hypothetical or tentative way; he is instead emphatic in insisting that our ignorance of many of the details of its operation does not justify any hesitation in accepting the fundamental structure of inheritance that it proposes:

> We cannot now fail to be impressed with the fallacy of reckoning inheritance in the usual way, from parents to offspring, using those words in their popular senses of visible personalities [i.e. persons]. The span of the true hereditary link connects, as I have already insisted upon, not the parent with the offspring, but the primary elements of the two, such as they existed in the newly impregnated ova, whence they were respectively developed. No valid excuse can be offered for not attending to this fact, on the ground of our ignorance of the variety and proportionate values of the primary elements: we do not mend matters in the least, but we gratuitously add confusion to our ignorance, by dealing with hereditary facts on the plan of ordinary pedigrees—namely, from the *persons* of the parents to those of their offspring. (BR 400–401, original emphasis)

The clearest evidence of Galton's failure to conceive of or consider important alternative theoretical explanations of hereditary phenomena is to be found in the precise manner in which he develops his case for the stirp theory, especially for his claim that certain of its commitments are simply unavoidable. He begins "A Theory of Heredity" by noting a consensus among his contemporaries in favor of Darwin's view (quoted from VAP v. 2 453) that "the body consists of a multitude of 'organic units', each of which possesses its own proper attributes, and is to a certain extent independent of all the others," and he insists that we may "rest assured" that this hypothesis "and all that such an hypothesis implies, must lie at the foundation of the science of heredity" (TH 329). Most importantly, he then goes on to describe what he calls "the four postulates that seem to be almost necessarily implied by any hypothesis of organic units" (TH 331):

> The first is, that each of the enormous number of quasi-independent units of which the body consists, has a separate origin, or germ. The second is, that the stirp contains a host of germs, much greater in number and variety than the organic units of the bodily structure that is about to be derived from them; so that comparatively few individuals out of the host of germs, achieve development. Thirdly, that the undeveloped germs retain their vitality: that they propagate themselves while still in a latent state, and contribute to form the stirps of the offspring. Fourthly, that organisation wholly depends on the mutual affinities and repulsions of the separate germs; first in their earliest stirpal stage, and subsequently during all the processes of their development.[14]

Concluding "A Theory of Heredity" Galton repeats that these postulates appear to be the "necessary consequences" (TH 347) of any effort to theorize about heredity on the basis of organic units, and in this recurring claim we find clear evidence of several respects in which he failed to appreciate important alternative theoretical possibilities.

Consider, for example, the first of Galton's postulates: that each of the "quasi-independent units,"[15] of which an organism's body consists is produced by a separate germ. This claim seems to ignore the possibility that inherited germinal units might produce the characteristics of organisms without any structural isomorphism between particular germinal and organic units, that is, without the transmission of a single discrete and isolable hereditary unit specific to each organic unit or constituent of an organism's body that "possesses its own proper attributes" (or as we might now say, without a distinct hereditary unit specific to each cell of the body, each tissue, or each phenotypic trait).

Consider also Galton's fourth consequence, that the organization or structure of an organism's body "wholly depends on the mutual affinities and repulsions of the separate germs," which would seem to ignore instead the possibility that the physical organization of the resulting body need not occur in virtue of a corresponding physical organization among the germinal materials themselves. Here Galton simply assumes that if inherited particulate germs are responsible for the organic units making up an organism's body, then a spatial or physical organization *among* those inherited germs must be the mechanism whereby the structure or organization of the resulting body is achieved. And this requirement is, of course, perspicuously violated by more recently influential conceptions of inherited particulate germinal materials since (at least) the rise of Mendelian genetics.

Indeed, these twin imaginative failures would themselves seem to be rooted in another that is still more fundamental: Galton seems to share with Darwin the idea that the effects of inherited germs on the resulting organism must be achieved by the germs themselves *growing into* or *becoming* the constituent units of that organism. This *maturational* conception of hereditary influence does indeed suggest (though perhaps not strictly require) that each trait, tissue, or physical constituent of the body must have a separate germinal source and that the organization or structure of the germinal materials must at some point reflect the organization or structure of the organism that is to be generated by them. This maturational conception ignores, however, the alternative possibility that germinal materials might instead causally influence the characteristics of an organism simply by directing or controlling the growth and/or development of the organism itself (or perhaps even exert causal influence in some other way altogether). And this alternative, *directive* conception of hereditary influence seems not even to suggest, much less to entail, either a one-to-one correspondence between germinal materials and constituent physiological elements of the body (Galton's first consequence) or the need for germinal materials themselves to exhibit at any point (whether by "mutual affinities and repulsions of the separate germs" or in any other way) the structure or physical organization that will ultimately emerge in the organism whose growth and development they direct (Galton's fourth consequence). Thus, Galton's confidence in the necessity of his first and fourth consequences seems rooted in his failure to conceive of even the possibility of any directive alternative to the maturational conception of particulate heredity he shared with Darwin.

Perhaps even more fundamentally, however, Galton's claim that any hypothesis of organic units requires the existence of many undeveloped germs which retain their vitality and are passed on to offspring (the second and third of his "necessary consequences" of any theory of organic units) illustrates his failure to consider any alternative to what we might call an *invariant* conception of heredity; that is, a conception on which each active or developed hereditary element exerts a specific, recognizable effect no matter what others are inherited along with it and no matter the context in which it occurs. By contrast, on a *contextual* conception of hereditary influence, hereditary materials might be present and fully 'developed' or active in the parents, but simply have different causal consequences in the context of a different suite of further inherited materials or against a different environmental background. Galton's failure to consider such contextual alternatives becomes especially clear in his explicit argument for the second necessary consequence:

> That the stirp contains a much greater variety of germs than achieve development, is proved by the fact that a person is capable of transmitting a variety of ancestral peculiarities to his children, that he did not himself possess. But since everything that reached him from his ancestors must have been packed in his own stirp, it follows that his stirp

> contained in addition to such peculiarities as were developed in his own bodily structure, those numerous other ancestral peculiarities of which he was personally destitute, but which he bequeathed to one or more of his descendants. Therefore, every stirp must be held to contain a great variety of germs in addition to those that may achieve development in the person who grows out of that stirp. (TH 331–332)

As this passage makes clear, Galton thinks that the phenomenon of reversion to ancestral characteristics simply requires any serious hypothesis of organic units to appeal to *latent* or *undeveloped* germinal materials (as both pangenesis and Galton's own stirp theory do). But this follows only on the invariant conception's assumption that an inherited germinal unit must always have precisely the same effect that it did in every ancestor when it "achieve[s] development" at all, or (in the language of "Blood Relationship") is 'patent' rather than 'latent'.[16]

What Galton is missing here, of course, is the contextualist possibility that hereditary materials might exist in offspring in just the same active and developed form that they did in ancestors, but produce different characteristics in different causal contexts. This sort of possibility clearly characterizes contemporary molecular genetics, of course, where differences in the phenotypic expression of a given gene are characteristically regarded as a matter of causal context and interaction, and even classical Mendelian relations of dominance and recessiveness seem most naturally considered as interactions between alleles rather than latency, dormancy, or some kind of undeveloped state of a gene itself.[17] Thus, Galton's confidence in the necessity of the second and third of his postulates seems to reflect his failure to conceive of the possibility of accounting for reversion equally well on a contextual theory of particulate material inheritance that assumes the intermittent recurrence of those (internal and/or external) causal conditions that enabled a particular germinal element to produce a particular characteristic in a given ancestor.

Once Galton's failure to conceive of these alternatives to his maturational and invariant conception of particulate heredity is made explicit in this way, it becomes relatively easy to identify further points at which they and the consequences that follow from them dominate his central writings on generation and inheritance. In the course of his attack on Darwin's pangenesis in the original version of "A Theory of Heredity," for instance, Galton argues that:

> [t]he germs that become developed into structure, are relatively too few to exert much hereditary influence, and when fully developed they would be somewhat passive and sterile. I argue, that as fertility resides somewhere, it must have been vested in the non-developed residue of the stirp, or rather in its progeny and representatives (whatever, or however numerous, they may be) at the time when the individual has reached adult life. (TH [1875] 88)[18]

Galton's failure to conceive of any alternative to the invariant conception of inheritance is here reflected in his insistence that the germs that will affect future generations must be in a "non-developed" state, while his failure to conceive of any alternative to the maturational conception is reflected in the notion that the formation of an organism is a matter of the inherited germs *themselves* becoming "developed into structure" and reaching a "fully developed" state.[19]

These same presumptions about particulate heredity appear throughout Galton's earlier and later writings about inheritance as well. In "On Blood Relationship," for instance, Galton argues that "because ancestral qualities indicated in early life frequently disappear and yield place to others," the organism "must receive supplementary contributions derived from their contemporary latent elements" (BR 396), but this follows only if we assume an invariant conception of what a single developed hereditary element can do.

Similarly, in "On Blood Relationship" we find Galton pursuing the following line of argument:

> From the well-known circumstance that an individual may transmit to his descendants ancestral qualities which he does not himself possess, we are assured that they could not have been altogether destroyed in him, but must have maintained their existence in a latent form. Therefore each individual may properly be conceived as consisting of two parts, one of which is latent and only known to us by its effects on his posterity, while the other is patent and constitutes the person manifest to our senses. (BR 394)

Most striking here is Galton's unselfconscious slide from the claim that recurring ancestral characteristics must somehow have persisted in a latent *form* to the idea that individual organisms must therefore be themselves made up of distinct patent and latent *parts*. Even if the notion of latent 'form' is sufficiently abstract to leave room for a contextual rather than invariant interpretation, Galton's further insistence that such latency must consist in undeveloped *parts* of ancestors transmitting those characteristics that they do not themselves express is a signal indication of his failure to conceive of any alternative to the invariant conception of hereditary influence.

Nor is this simply a momentary slip or oversight on Galton's part: he consistently treats the patent and latent elements as disjoint sets of material components, claiming, for example, that "the latent elements must be greatly more varied than those that are personal or patent" (BR 395) and that it follows from his arguments "that for each place among the personal elements there may exist, and probably often does exist, a great variety of latent elements that formerly competed to fill it" (BR 395). And in the much later work *Natural Inheritance* (hereafter NI), he again argues that:

> The existence in some latent form of an unused portion is proved by his power, already alluded to, of transmitting ancestral characters that he did not personally exhibit. Therefore the organised structure of each individual. . . . is the coherent and more or less stable development of what is no more than an imperfect sample of a large variety of elements." (18/318)

This passage illustrates, just as Galton's earlier writings did, not only the invariant conception's insistence that distant reversion requires distinct latent and patent parts of the organism, but also the maturational conception's insistence that inherited germs themselves develop into or become the various parts of an organism's body, captured here in the insistence that "the organised structure of each individual. . . . is the coherent and more or less stable development of" a sample of the inherited germs themselves (18/318). Galton's exclusive commitment to the maturational conception is also widely represented in "On Blood Relationship," where he writes that "The embryonic elements are *developed* into the adult person" (395, original emphasis) and that "the embryonic elements are . . . *developed (a)* into the visible adult individual" (396, original emphasis). Indeed, in the course of this discussion he goes so far as to coin distinct terms for two different processes of development to distinguish the process whereby a latent germ ultimately becomes part of the stirp that will be passed along to the next generation ("development (b)") from the process by whereby a patent germ becomes part of the body of the adult organism ("development (a)").[20]

It is worth noting that Galton's failure to recognize the possibility of directive and/or contextual alternatives to his maturational and invariant conception of heredity (and to the consequences he draws from it) persists even when he is explicitly concerned to emphasize what (and how much) remains to be discovered about the processes of inheritance. He writes, for instance, that the principle on which "the embryonic elements are segregated from among [the stirp]":

> may be described as . . . "*Class Representation*," using that phrase in a perfectly general sense as indicating a mere fact, and avoiding any hypothesis or affirmation on points of detail, about most, if not all, of which we are profoundly ignorant. I give as broad a meaning to the expression as a politician would give to the kindred one, a "representative assembly." By this he means to say that the assembly consists of representatives from various constituencies, which is a distinct piece of information so far as it goes, and is a useful one, although it deals with no matter of detail. . . . (BR 395)

Galton allows that in the case of heredity we are quite ignorant about many aspects of the corresponding "elections," including the stirpal analogues of the "number of electors," their "qualifications," their "motives," the "number of seats," "how many candidates there are usually for each seat," and any number of further details, but once all this is conceded, he argues, "there can, I trust, be no difficulty in accepting my definition of the general character of the relation between the embryonic and structureless elements, that the former are the result of election from the latter on some method of Class Representation" (BR 395). Thus, even in those cautious moments when Galton is explicitly concerned with our remaining ignorance about aspects of heredity, it does not occur to him to question the invariant conception's insistence that an organism must consist of distinct patent and latent component parts.[21]

Galton assures us that the maturational conception's language of "development" is also carefully calculated to allow for our remaining ignorance about inheritance, for:

> "Development" is a word whose meaning is quite as distinct in respect to form, and as vague in respect to detail, as ["Class Representation"]; it embraces the combined effects of growth and multiplication, as well as those of modification in quality and proportion, under both internal and external conditions." (BR 395–396)

Galton's discussion here and throughout (see above), however, does not simply concern the development of the organism, but instead the development *of* the inherited germs *into* the organism. This process, no matter how "vague in respect to detail," would seem to require (as the maturational conception insists) that the inherited germs themselves ultimately *become* the constituent parts of the adult organism, and indeed, Galton elsewhere insists that "growth, nutrition, and reproduction . . . are all due to the development of the same germinal matter, variously located" (TH 343–344). Galton also embraces this implication when he goes on to assure us, for instance, that an approximate knowledge of the original elements "would no doubt enable us to predict the average value of the form *into which they would become developed*" (BR 396, my emphasis).

In a similar fashion, Galton presumes that an organism's structure must be produced by a corresponding physical configuration among the inherited germs even when he goes out of his way in both "A Theory of Heredity" and *Natural Inheritance* to concede our complete ignorance of the "repulsions and affinities" among germs whereby that configuration is effected:

> It is certain, from the rapidity of the visible changes in the substance of the newly fertilized ovum, that the germs in the stirp are in eager and restless pursuit of new positions of organic equilibrium. . . . We know nothing yet of the nature of these repulsions and affinities. . . . but we ought . . . to expect them to act on many sides, in a space of three dimensions, just as the personal likings and dislikings of an individual in a flying swarm may be supposed to determine the position that he occupies in it. . . . We may therefore feel assured that the germs must be

> affected by numerous forces on all sides, varying with their change of place, and that they must fall into many positions of temporary and transient equilibrium, and undergo a long period of restless unsettlement, before they severally attain the positions for which they are finally best suited. (TH 335; cf. NI 20/320–21/321)

Here again Galton's point is to acknowledge what he allows is our considerable remaining ignorance about the processes of inheritance, but it never occurs to him to question the maturational conception's assumption that the inherited germs must take on new and ever more complex structural configurations *among themselves* in order to produce a fully developed organism with a corresponding physical structure. Once again, then, Galton presupposes that particulate inheritance must be maturational even when he addresses the aspects of heredity concerning which he concedes our complete ignorance and most explicitly seeks to leave room for a variety of alternative possibilities.

4.3 Galton's Understanding of "Correlation" and "Variable Influences" in Development

Curiously, however, Galton's acknowledgement of our remaining ignorance about heredity does at one point at least seem to suddenly recognize the possibility of contextual rather than invariant hereditary influence, for he writes that we are also ignorant of "whether the result of the elections at one place may or may not influence those at another (on the principle of correlation)" (BR 395). But his later elaboration of this "principle of correlation" makes clear that this concession is in no way intended to leave room for a contextual alternative to the invariant conception. He says:

> Lastly, I must guard myself against the objection that though structure is largely correlated, I have treated it too much as consisting of separate elements. To this I answer, first, that in describing how the embryonic are derived from the structureless elements, I expressly left room for a small degree of correlation; secondly, that in the development of the adult elements from the embryonic there is a perfectly open field for natural selection, which is the agency by which correlation is mainly established; and thirdly, that correlation affects groups of elements rather than the complete person, as is proved by the frequent occurrence of small groups of persistent peculiarities, which do not affect the rest of the organism, so far as we know, in any way whatever. (BR 396)

As this passage suggests, Galton's confession of ignorance concerning whether the outcome of one "election" might not influence another on the principle of correlation is intended to leave room simply for the possibility that the selection of one germ for the embryo might affect which *other germs* are also *chosen* to become "patent" or developed and not for the genuinely contextualist possibility that the development of one germ might influence what another germ would become, produce, or grow into. In *Natural Inheritance*, Galton clarifies the sort of correlation he allows for with an elaborate analogy:

> Suppose we were building a house with second-hand materials carted from a dealer's yard, we should often find considerable portions of the same old houses to be still grouped together. Materials derived from various structures might have been moved and much shuffled together in the yard, yet pieces from the same source would frequently remain in juxtaposition and it may be entangled. They would lie side by side ready to be carted away at the same time and to be re-erected together anew. So in the process of transmission by inheritance, elements derived from the same ancestor are apt to appear in large groups, just as if they had clung together in the pre-embryonic stage, as perhaps they did. They form what is well expressed by the word "traits," traits of feature and character—that is to say, continuous features and not isolated points. (NI 8/308–9/309)

Room for a similar kind of misunderstanding arises in the course of Galton's argument that from "an approximate knowledge of the original elements" we would be able "to predict the average value of the form into which they would become developed," for he there allows that "the individual variation of each case would of course be great, owing to the large number of variable influences concerned in the process of development" (BR 396). But again Galton's later and more elaborate discussion of this process in *Natural Inheritance* helps to make clear why this concession cannot reasonably be read as recognizing the possibility of a contextual rather than invariant account of hereditary influence:

> It would seem that while the embryo is developing itself, the particles more or less qualified for each new post wait as it were in competition, to obtain it. Also that the particle that succeeds must owe its success partly to accident of position and partly to being better qualified than any equally well placed competitor to gain a lodgment. Thus the step-by-step development of the embryo cannot fail to be influenced by an incalculable number of small and mostly unknown circumstances. (NI 9/309)

Thus, while Galton recognizes a "large number of variable influences" and "an incalculable number of . . . unknown circumstances" at work in development, he consistently sees these influences simply as affecting *which germs are chosen* to become the patent organism rather than the form into which those germs might develop once selected or the character of that development. And, of course, whenever Galton seeks to explain the differences between related individuals, whether parents and their children, siblings, or even twins, he invariably appeals to differences between the respective collections of germs inherited or selected for development[22] (or, much more rarely, modifications of the germs themselves prior to their inheritance and/or development, e.g. TH 344) rather than to interactions between germs or differences in the causal context of their development itself (see, e.g., BR 400–402; TH 336–338, 339–340, 343–345; NI 7/307–12/312). Perhaps most significant of all, Galton's explicit and apparently exhaustive discussion of the sources of individual variation in "A Theory of Heredity" limits the influence of external circumstances in precisely this same way:

> Individual variation depends upon two factors; the one, is the variability of the germ and of its progeny; the other, is that of all kinds of external circumstances, in determining which out of many competing germs, of nearly equal suitability, shall be the one that becomes developed. (TH 338)

Thus, Galton simply does not recognize any *sources* of hereditary variation and individual difference that leave room for contextual alternatives to the invariant conception of particulate heredity.

Indeed, Galton's failure to recognize the possibility of a contextual account of heredity is especially telling in light of the fact that he clearly did recognize that what we ordinarily regard as single traits of organisms might well involve the action of multiple, independent, hereditary germs. Arguing that the stirp must have limited space for germs of each variety, for instance, he claims that:

> ... in the gradual breeding-out of negro blood, we may find the colour of a mulatto to be the half and that of a quadroon to be the quarter of that of his black ancestors; but as we proceed further, the subdivision is very irregular; it does not continue indefinitely in the geometrical series of one-eighth, one-sixteenth, and so on, but it is usually present very obviously, or not at all, until it entirely disappears. There are many more gradations in compound results, as in an expression of the face, because any one of its elementary causes may be present or absent; and as the number of possible combinations or alternatives, among even a few elements, is very great, there must be room for a large number of grades between the complete inheritance of the expression and its total extinction. (TH 335)

Because Galton here explicitly recognizes that identifiable traits like skin color or "an expression of the face" may involve the action of several different inherited germs, it is all the more striking that this recognition is limited to the possibility that organismic traits or features might involve *combining* the developed products of multiple hereditary germs (the "combination" of "elementary causes" into a "compound result") rather than the genuinely contextualist possibility that the presence or absence of one hereditary germ might affect the character or characteristics of the physiological unit that another such germ would become or produce. Galton contrasts such combinatorial action only with cases in which the "slow loss of some characteristic of a race" is mediated by changes in the germs themselves and cases in which one developed organismic *quality* has simply "overpowered" another (TH 334). Thus, even when Galton is most sensitive to the question of what interactions among hereditary germs might be required to produce characteristics of organisms themselves, he shows no evidence of recognizing the possibility that what one inherited germ itself does or develops into might depend upon the other germs that have been inherited or indeed on any other aspect of its causal context.

But doesn't the very analysis Galton offers above of the inheritance of skin color in the case of "negro blood" run counter to both the maturational and the invariant conception of particulate inheritance? After all, on Galton's account it seems that the intermediate color of the skin cells of a mulatto or quadroon is achieved by a small number of germs (dwindling eventually to just one, "present very obviously or not at all") affecting the color of all those *others* which develop into skin cells, and also that the inheritance of at least some discrete germs thus affect the characteristics of the resulting organism in some way besides developing *into* one of its constituent parts.[23] How, then, could Galton have offered this suggestion about the inheritance of skin color without recognizing alternatives to both the maturational and invariant conceptions of inheritance?

That Galton's analysis of this case seems to modern eyes to *demand* alternatives to his maturational and invariant conception of heredity makes it especially revealing that Galton himself did not see matters in this way. In his exchange of correspondence with Galton concerning "A Theory of Heredity," Darwin wrote to say that although he was prepared to admit that gemmules might be largely multiplied in the sex organs, "this does not make me doubt that each unit of the whole system also sends forth its gemmules" and to ask for Galton's response to the following objection:

> If 2 plants are crossed, it often or rather generally happens that every part of the stem, leaf—even to the hairs—and flowers of the hybrid are intermediate in character; and this hybrid will produce by buds millions on millions of other buds all exactly reproducing the intermediate character. I cannot doubt that every unit of the hybrid is hybridised and sends forth hybridised gemmules. Here we have nothing to do with the reproductive organs. (Dec. 18, 1875; LLL II 189)

Galton replied that "The explanation of what you propose does not seem to me in any way different on my theory to what it would be on any theory of organic molecules" (Dec. 19, 1875; LLL II 189). He goes on to argue that in such a case an intermediate character like the gray tint of a hybrid midway between its black and white parents would actually be produced by a pattern of discrete individual white and black cells (or smaller physiological constituents) developed from germs with distinct origins. That is, Galton insisted that the germs themselves need not be hybridized in order to produce the hybridized trait, for the color of the innumerable plants produced by budding might be achieved not by inheriting gray (hybridized) gemmules at all, but instead by inheriting a given *proportion* of white and black gemmules.[24] Galton actually proposes two alternative versions of this scenario (both illustrated in his response to Darwin's letter). If the cells were what he calls both "structural" and "organic" units, the organism under sufficient magnification would exhibit a mosaic pattern of distinct black and white cells. If instead the organism's cells were themselves gray, this would imply that the fundamental structural unit of the body (the cell) was not the same as its fundamental organic unit: each cell would be instead an "organic molecule" with discrete black and white gemmules (the individual organic units) both represented within it. Indeed, Galton goes on to propose that on the second account the number of grades of variation in which such a trait could appear would provide a way to determine how many gemmules were involved in constituting each organic molecule (structural unit) of the body:

> It has been an old idea of mine, not yet discarded and not yet worked out, that the number of units in each molecule may admit of being discovered by noting the relative number of cases of each grade of deviation from the mean greyness. If there were 2 gemmules only, each of which might be either white or black, then in a large number of cases one-quarter would always be quite white, one-quarter quite black, and one-half would be grey. If there were three molecules [Galton must mean 'gemmules' here], we should have 4 grades of colour (1 quite white, 3 light grey, 3 dark grey, 1 quite black and so on according to the successive lines of "Pascal's triangle"). (Dec. 19, 1875; LLL II 190)[25]

What is striking here and throughout is Galton's insistence that variations of color in such hybrids are produced by a mixture or mosaic pattern of locally invariant, maturational hereditary units too fine to be resolved individually, along with his assurance that this is the explanation "any theory of organic units" would give! On his account, all individual gemmules would produce *only* white or black organic units (reflecting the invariant conception of heredity), whether these organic units are themselves the structural units of the body (e.g. cells) or simply the constituents of such structural units. Likewise, an organism would exhibit a trait like gray color at the organismic level not in virtue of having inherited one or a few hereditary germs with an effect on the organism as a whole, but instead only in virtue of the *proportions* of individually invariant hereditary particles that had become developed into a patent form (reflecting the maturational conception).

We must consider the possibility, however, that Galton was forced into this analysis by the fact that Darwin's objection concerned hybrids reproduced by budding and that he did not intend it to be general in scope. How do we know that Galton conceived of the inheritance of "negro blood" in this way as well? In fact, Galton returns to the inheritance of skin color in *Natural Inheritance* and his discussion there makes clear that his analysis of the gray color of an intermediate hybrid is the explanation he envisions for this case as well:

> As regards heritages that blend in the offspring, let us take the case of human skin colour. The children of the white and the negro are of a blended tint; they are neither wholly white nor wholly black, neither are they piebald, but of a fairly uniform mulatto brown. The quadroon child of the mulatto and the white has a quarter tint; some of the children may be altogether darker or lighter than the rest, but they are not piebald. Skin-colour is therefore a good example of what I call blended inheritance. It need be none the less "particulate" in its origin, but the result may be regarded as a fine mosaic too minute for its elements to be distinguished in a general view." (NI, 12/312)

This passage seems to put beyond question that Galton thought of the inheritance of negro blood along the same lines he proposed in his answer to Darwin, and thus that he did not see his own analysis as even inviting (much less requiring) directive and/or contextual alternatives to his own maturational and invariant conception of heredity.

A final question must be faced. If skin color is simply a matter of the proportion of maturational, invariant germs inherited from ancestors, why should it be that the inheritance of skin color "would not continue indefinitely in the geometrical series of one-eighth, one-sixteenth, and so on, but it is usually present very obviously, or not at all, until it entirely disappears"? Of course, this is easy to understand if there are a limited number of places for organic units in the organic molecules constituting the structural units of the body: the lightest grade of color would simply correspond to having only one black gemmule in each organic molecule, the minimum proportion on which any color will be produced at all. But if instead the cells of the body are themselves both the organic and structural units (that is if each gemmule develops into a cell of the body), why should there be any limit on the proportion of cells that could be darkened? Here Galton's explanation would have to appeal to the principle of correlation discussed above, and insist that a minimum proportion of gemmules destined to develop into darkened cells would have to be inherited all together or not at all. But even on this account of the matter Galton preserves his evidence for limited space in the stirp, for if the stirp could increase without measure (i.e., there were an ever-increasing amount of room for gemmules of each type in each stirp), any minimum *number* of color-producing gemmules that must be inherited together or not at all would not constitute a minimum *proportion* at all, for the proportion of the stirp represented by such a minimum number of gemmules would continually decrease as the total number of gemmules in the stirp increased. Thus, Galton is well within his rights to appeal to the case of negro blood to argue for limitations of space in the stirp even on the maturational, invariant analysis he seems unable to see past and even if we attribute (perhaps implausibly) to him at the time he used it in this way the fully detailed analysis of hybrid color he would later offer in response to Darwin's challenge.

Despite some first appearances, then, a detailed exploration of Galton's understanding of correlation in inheritance and variation in development turns out simply to provide striking further evidence for the same conclusion supported more generally and explicitly throughout Galton's writings on inheritance and generation: that he never conceived of the possibility of directive or contextual accounts of particulate inheritance, or indeed of any alternatives to the maturational and invariant aspects of his own conception, despite the fact that the phenomena to which he appealed supported a directive and/or contextual version of *his own theory* equally well. And of course it is not simply *particular* alternative theories of inheritance (such as Mendelian genetics or contemporary molecular genetics) whose very possibility went unrecognized by Galton, but instead entire broad *classes* of such alternatives: that is, directive or contextual theories quite generally. There seems little room to doubt the scientific seriousness of such alternative views of particulate inheritance, which would, after all, be accepted by the scientific community not long after Galton's own account was developed. Instead, it seems we must conclude that just as Darwin failed to conceive of the very possibility of any common-cause mechanism for inheritance, after surmounting this conceptual obstacle Galton failed in turn to conceive of any alternatives to the maturational and invariant aspects of his own account of particulate inheritance. But further significant developments were in store before the close of the century, and it is to the great conceptual innovations and equally profound imaginative failures of August Weismann that we now turn our attention.

Notes

1. Independently conceived by G. J. Romanes (Cowan 1985 110n).

2. In addition to the direct injection of blood (stirred to "defibrinize" it) from one rabbit to another, Galton later successfully maintained the cross-circulation of blood between the carotid arteries of different rabbits (connected by means of cannulae) for as long as 37 and one-half minutes, producing, by his estimate, rabbits with fully one-half "alien" blood (see Robinson 1979 34). It appears that he later also tried repeating the experimental transfusions for multiple generations in a given lineage of rabbits (LLL II 156ff). The negative outcome of the latter experiments is discussed briefly in "A Theory of Heredity" (1876 341–342).

3. And indeed, in later editions of VAP Darwin modified the two passages Galton noted: he replaced the expression "circulate freely" with "are dispersed

throughout the whole system" and dropped out altogether his claim that the thorough diffusion of gemmules was not unlikely given "the steady circulation of fluids throughout the body" (see LLL II 164n).

4. Darwin's further interest in these experiments is somewhat puzzling, as his letter to *Nature* suggested that they had no bearing on pangenesis: he not only pointed out that blood cannot be the medium for the circulation of gemmules in Protozoa or plants, but also argued that "the means by which the gemmules... are diffused through the body, would probably be the same in all beings; therefore the means can hardly be diffusion through the blood" (cited in LLL II 163). As Pearson suggests (LLL II 165–166), it is hard to resist the suspicion that Darwin thought some evidence in favor of pangenesis might still be forthcoming from the transfusion experiments nonetheless. None was.

5. This impression is conveyed even more strongly by Darwin's noticeably less shrill reply to the transfusion experiments in the second edition of VAP: "I certainly should have expected that gemmules would have been present in the blood, but this is no necessary part of the hypothesis, which manifestly applies to plants and the lowest animals" (1876 II 350). Interestingly, Darwin here seems to regard it as having been reasonable to expect gemmules to circulate in the blood despite the fact that this almost flatly contradicts his suggestion in the letter to *Nature* (above) that the mechanism for diffusing gemmules is probably the same in all organisms.

6. We might say instead that the presumption about circulation was part of pangenesis itself and that Darwin modified the theory under pressure from Galton's experiments, but the holist's claim about the open-ended character of such available modifications remains unchanged.

7. We cannot, of course, eliminate all the *unserious* or wildly skeptical avenues of response to experimental disconfirmation (pleading hallucination, conspiracy theories, and so forth), nor can the very standards we use to judge such possibilities to be unserious be defended *ex nihilo*. But if these are supposed to be our reasons for concern then the holist is also open to the now-familiar charge that she has merely repackaged familiar, unanswerable Cartesian skepticism and offers no special cause for concern about theoretical scientific beliefs that does not apply with equal force to all beliefs whatsoever.

8. Some sense of Galton's reverence for Darwin can be gained not only from his response to Darwin's letter to *Nature* concerning the transfusion experiments (see above), but also from the following letter written to his sister, Emma Galton, two days after Darwin's death: "Dearest Emma, I feel at times quite sickened at the loss of Charles Darwin. I owed more to him than to any man living or dead; and I never entered his presence without feeling as a man in the presence of a beloved sovereign. He was so wholly free of petty faults, so royally minded, so helpful and sympathetic. It is a rare privilege to have known such a man, who stands head and shoulders above his contemporaries in the science of observation.... The world seems so blank to me now Charles Darwin is gone. I reverenced and loved him thoroughly. Ever affectionately, Francis Galton" (LLL II 198). Galton's devotion to Darwin was equally fervent and sincere in public, as witnessed by his speech to the Royal Society in 1886 (four years after Darwin's death) upon receiving its gold medal: "Few can have been more profoundly influenced than I was by his publications. They enlarged the horizon of my ideas. I drew from them the breath of a fuller scientific life, and I owe more of my later scientific impulses to

the influence of Charles Darwin than I can easily express. I rarely approached his genial presence without an almost overwhelming sense of devotion and reverence, and I valued his encouragement and approbation more, perhaps, than that of the whole world besides" (LLL II 201).

9. Of course this implies that Darwin's proposal does not meet the strict standard of scientific seriousness we are employing generally (viz. acceptance by some actual scientific community), but the idea that gemmules could be diffused directly through tissues nonetheless seems clearly not to be an unserious possibility in the sense of Cartesian Evil Demons or pleading hallucination.

10. "A Theory of Heredity" originally appeared in the December 1875 issue of the *Contemporary Review*, but was substantially revised for its publication in the 1876 volume of the *Journal of the Royal Anthropological Institute of Great Britain and Ireland*. Except where otherwise specified, all references in the text are to the latter, revised, version.

11. Galton allows, however, that a few such latent elements must become developed over the course of an organism's life "because ancestral qualities indicated in early life frequently disappear and yield place to others" (BR 396).

12. It is extremely difficult to understand the contrast Galton means to draw between "Class Representation" and "Family Representation" (cf. LLL II 170–172). Cowan suggests that in Family Representation "representative determinants from each familial strain would be selected" (124), but this suggestion seems undercut by Galton's remark that "the similar elements contributed by the two parents rank[], of course, as of the same family" (BR 397). He sometimes writes as if families are *defined* as larger groupings made up of classes, but later claims that it is only "most probable" that Family Representation "is in reality a large selection made out of larger and not out of smaller constituencies than those I have called 'classes'" (BR 397). One salient difference seems to be that Class Representation selects only a single germ for development into a given unit of the body, while Family Representation selects an entire group of corresponding hereditary elements for inclusion in the stirp. At least some of the representational power of this latter process, however, (whether of gemmules drawn from the various ancestral lineages or corresponding to the various parts of the body) is supposed to be ensured by statistical distribution alone, as Galton analogizes selection by Family Representation to the formation of an army by general conscription, noting that the constitution of such an army "according to the laws of chance, will reflect with surprising precision the qualities of the population whence it was taken; each village will be found to furnish a contingent, and the composition of the army will be sensibly the same as if it had been due to a system of immediate representation from the several villages" (BR 397–398).

13. The earlier version reads "wholly supplementary and subordinate" (TH [1875] 81).

14. Immediately following his discussion of these "postulates," Galton remarks "We must also bear in mind that the alternative hypothesis of a general plastic force resembles that of other mystic conceptions current in the early stages of many branches of physical science, all of which yielded to molecular views, as knowledge increased" (TH 332). Thus, Galton seems to regard the hypothesis of "a general plastic force" as the only serious alternative to "molecular" views of inheritance, which in turn he thinks must be characterized by the four postulates he describes.

15. To support the first consequence Galton first argues that "the independent origin of the several parts of the body may be argued from the separate inheritance of their peculiarities" (TH 331), and the illustrative example he offers is a child who inherits its father's eyes and its mother's mouth. But as we will see later in this chapter, elsewhere he clearly countenances the idea that each particular germ might be individually responsible for producing only a single cell of the body or even just a constituent element of one of those cells.

16. Elsewhere (TH 343) Galton describes the gemmules that do not achieve development in a given individual as "dormant."

17. The closest analogue in contemporary molecular genetics to Galton's latency or undeveloped state of inherited particulate germinal material would be a gene's failure to be "turned on" by the appropriate regulatory gene, but even this would seem to be a matter of *interaction* between inherited germinal units and not a difference of *development* between them: indeed, the action of regulatory genes *constitutes* one of the ways in which inherited particulate germinal materials can fail to do what they did in an ancestor without themselves being inherited in a latent state or undeveloped form at all. And of course, many other differences in the phenotypic expression of a gene are not caused by differences in the operation of such regulatory genetic machinery in any case.

18. An expanded version of this argument appears in the revised version of TH (1876 339).

19. Note, however, that Galton's insistence that fully developed germs would be sterile or passive is *not* a feature of the maturational conception of hereditary influence he shared with Darwin. That developed germs were themselves the source of further hereditary particles was, of course, the central contention of pangenesis; Galton was well aware of this possibility and indeed sought to refute it. It is the idea that inherited germs exert their influence by themselves *growing into or becoming* the organic units, tissues, or other physical constituents of an organism's body to which neither Darwin nor Galton seems to have recognized any particulate alternative.

20. Galton's diagram (BR 398, Fig. 1) explicitly specifies that the embryonic elements (selected by "Class Representation" from the inherited stirp) "become" the component parts of the body of the adult organism by the process of development (a), just as the latent elements in the embryo "become" the latent elements in the adult by development (b).

21. Galton makes a similar claim about the corresponding selective process of "Family Representation," arguing that despite our ignorance on matters of detail, "It is most important to bear in mind that this phrase states a fact and not an hypothesis. . . . " (BR 397). But such a claim about *Family* Representation (in which only some of an organism's latent gemmules are selected for the stirp that will be passed on to its offspring) would not seem to reflect any failure to imagine alternatives to either a specifically maturational or invariant conception of heredity.

22. As the passages above illustrate, when Galton's explanation of the similarity of "true" twins appeals in part to "the circumstances under which the bodily structure is developed out of them being almost identical" (TH 337), he is also recognizing a role for external circumstances in selecting particular germs for development, rather than recognizing the possibility of a contextual alternative to his invariant conception of heredity.

23. Even if Galton is still free to require that each physiological unit of the body must develop from a corresponding hereditary germ, he might seem here to allow that not all germs must develop into such physiological units.

24. This line of thought is present in Galton's writing in at least an embryonic form as far back as "On Blood Relationship." In explaining variation from hybrid characters in the offspring of a true hybrid, Galton writes, "A white parent necessarily contributes white elements to the structureless stage of his offspring, and a black, black; but it does not in the least follow that the contributions from a true mulatto must be truly mulatto" (BR 402). Of course, this does not help explain how a full complement of productive germs capable of reproducing the hybrid character could have reached the bud in the first place or how this process could be repeated indefinitely, which seems to have been a further aspect of Darwin's concern (cf. above and Cowan, 1985 264).

25. This famous passage in Galton's letter to Darwin led some commentators (most notably Olby in the first edition of his *Origins of Mendelism* (published in 1966) to suggest that Galton here anticipates Mendel's own explanation of inheritance, including the Mendelian ratios! As Cowan points out (1985 Appendix), however, Galton's were ratios *of* something entirely different (viz. phenotypes rather than genotypes). And Olby retracts this claim in the later edition (1985).

5

August Weismann's Theory of the Germ-Plasm

> There is no doubt that several of the geniuses to whom we owe modern physics have built their theories in the hope of giving an explanation of natural phenomena, and that some even have believed they had gotten hold of this explanation.... Chimerical hopes may have incited admirable discoveries without these discoveries embodying the chimeras which gave birth to them. Bold explorations which have contributed greatly to the progress of geography are due to adventurers who were looking for the golden land—that is not a sufficient reason for inscribing "El Dorado" on our maps of the globe.
>
> —Pierre Duhem, *The Aim and Structure of Physical Theory*

5.1 German Biology at the End of the Nineteenth Century and Weismann's Theory of the Germ-Plasm

In the latter decades of the nineteenth century, August Weismann faced a very different context of theorizing about generation and inheritance than had any of his illustrious predecessors. Perhaps most importantly, a growing program of research in Germany had followed Ernst Haeckel's early speculation (1866) that the nucleus of the cell was the bearer of hereditary material, and a new generation of microscopical anatomists including Edward Strasburger, Oscar Hertwig, and Weismann himself had turned their attention in earnest to examining the behavior of the nucleus and chromosomes.[1] Throughout the 1870's and beyond, improved instruments as well as newly developed techniques of immersion, fixation, and staining had rewarded this sustained attention with a rapid pace of landmark discoveries in nuclear cytology. In 1873 Anton Schneider observed the successive stages of mitosis (ordinary cell division, then called "indirect" cell division), including the progression of forms taken on by the nuclear elements.[2] In 1875 Hertwig witnessed fertilization in sea urchin eggs, noting that a single spermatozoon enters the ovum and its nucleus joins with that of the egg.[3] In 1879 Walther Flemming published his observations of the longitudinal splitting of the "chromatin threads" (chromosomes) during mitosis. And in 1883 Edouard van Beneden reported that the process of meiosis (the production of haploid sex cells by an initial doubling of the chromosomes followed by quadruple cell-division) involves a "maturation division" (later termed "reduction division" by Weismann) producing cells with only half the number of chromosomes found in a nonreproductive cell of the same organism. These findings and others generated new questions even as they suggested possible answers and galvanized interest in the study of hereditary transmission as such i.e., distinct from questions about the mechanics of growth, reproduction, and generation more generally).[4] Between 1883 and 1885, prominent German cytologists and experimental embryologists proposed no fewer than five different theories of inheritance grounded in the notion of some kind of continuity of material structure, not simply from parents to offspring but also within a single organism from early embryogeny to the production of gametes (see Churchill 1987).

One of these was an early version of Weismann's own account, presented in his famous inaugural address to the University of Freiburg in 1883 and published later that same year as the essay "On Heredity." It was in this address that Weismann himself first proposed (as Galton had before him) that an organism's germinal materials are sequestered from the beginning of its development and passed directly to that organism's offspring without being affected by events during the course of its life (except for the mixing of such materials required during sexual reproduction).[5] He had not by this point clearly formulated the concept of a germ-plasm, however, and so expressed this germinal continuity as a continuity of the reproductive *cells* from each generation to the next. Weismann conceived of the emergence of multicellular organisms as analogous to the formation of a colony of single-celled organisms, with a division of labor among cells specialized for different functions: most importantly, the germ cells retained the potential immortality of their single-celled ancestors (which had reproduced by simple division) while the cells of the rest of the body (the soma) became simply the evolutionary host and vehicle by which these germ cells were transmitted from one generation to the next. Thus, Weismann's initial formulation of germinal continuity proposed simply that an organism's germ cells are produced by division from its own early embryonic cells and sequestered before the development of its tissues and organs, an arrangement suggested to him by his own earlier work in experimental embryology (see Coleman 1965 153, Robinson 1979 153–154, Churchill 1986). Further evidence and criticism would ultimately force him to abandon this position and argue instead that only a germ-*plasm* or hereditary *substance* is reserved from the beginning of ontogeny and passed through a particular series of intervening somatic cells to the gametes[6] (1885; for discussion see Churchill 1987 346–347, 352–354). And Weismann's attention was drawn specifically to the nucleus when he became aware of van Beneden's work and realized that the behavior of chromosomes during reduction division and fertilization fit his own theoretical predictions concerning how the germ-plasm must be transmitted (see Bowler 1989 88).[7]

Weismann's ideas about inheritance and its material foundation, as well as the sequestration of the germ-line, would evolve and develop considerably between 1883 and 1892, the year in which he published his revolutionary landmark: *The Germ-Plasm, A Theory of Heredity* (English translation 1893; hereafter GP). Although this ongoing development would play an important part in undermining the very practice of theorizing about "generation"—in which the study of hereditary transmission was treated simply as a secondary aspect of the supposedly more fundamental question of how a new organism is manufactured or produced from the material of its parents' bodies (cf. chap. 3 and Bowler 1989 chaps. 2–4)—the fully mature theory Weismann himself produced was nonetheless remarkable for its comprehensive and synthetic character: it sought to account for the phenomena of growth, development, reproduction, regeneration, and ontogenetic differentiation in addition to those of hereditary transmission.[8] The impact of this mature theory would also ultimately prove to be the undoing of the then-dominant developmentalist tradition of theorizing about these questions, which conceived of heredity by means of a vague analogy with memory and treated embryological development itself as a model for a quite general tendency in nature for vital or teleological forces to push natural processes to ascend toward higher levels of complexity, organization, or perfection (see Bowler 1989 chap. 2). This developmentalist perspective was perhaps most famously captured in the claim of Haeckel's "biogenetic law" that in the course of its embryological development each individual organism is successively conveyed through the *adult* forms of the various preceding organisms in the evolutionary history of its species and thus that "ontogeny recapitulates phylogeny." Ironically, however, historians of the period have argued convincingly that the development of Weismann's own account relied on the notion of ontogenetic recapitulation and was thus itself firmly grounded within the developmentalist tradition whose doom it foretold (Churchill 1986 passim, Bowler 1989 87; Gould 1977 102–109). Moreover, the emergence of Weismann's distinctive account of inheritance and development from this foundation was itself an intricate and piecemeal process, as he came only slowly to reject Haeckel's conception of reproduction as a kind of "overgrowth" or excess production of the parent's own tissues,[9] as well as Schwann and Haeckel's reductionist conception of such growth as a process analogous to the formation of inorganic crystals by accretion, not to mention his own earlier steadfast commitment to epigenesis (see Coleman 1965 151–154, and esp. Churchill 1968). In their place he came to embrace the idea that development and inheritance would have to be explained by the transmission of discrete nuclear elements in a continuous germ line from ancestors to offspring, consisting of fundamental vital particles whose hereditary influence on an individual organism was somehow predetermined by and encoded in their respective heterogeneous material constitutions.

An important part of what prompted Weismann to develop the clear and comprehensive statement of his views offered in *The Germ-Plasm* was the publication in 1889 of a brief, challenging work on inheritance and generation entitled *Intracellular Pangenesis* by the botanist Hugo de Vries. Although Weismann embraced central elements of de Vries's account and it may have precipitated his own final break with epigenesis (see Churchill 1968 105–106), Weismann was also concerned in *The Germ-Plasm* to clarify crucial outstanding differences between his own position and the one de Vries proposed and to defend his own commitments on these matters (see Robinson 1979 chap. 8). Among these differences were de Vries's denial of the need for a reduction division and of any important distinction between somatic and germ tracks (as well as his related belief in the totipotency of all or most somatic cells), but perhaps most important of all was Weismann's insistence that the hereditary material must exhibit a precise hierarchical internal *structure* (as against de Vries's conception of his "pangenes" as freely mixing independent entities; see GP chap. 1 Sec. 4, esp. 69). On Weismann's account, the germ-plasm as a whole consisted of smaller constituent units called *idants*, which Weismann identified (probably but not certainly) with the chromosomes visible under the microscope. These idants were themselves composed of *ids*, each of which contained individually a sufficient amount and variety of hereditary material to produce a complete organism of the relevant species.[10] The ids were in turn made up of *determinants*, each responsible for directing the development and operation of either a single cell or multiple cells of a single uniform kind.[11] Finally, these determinants were themselves composed of Weismann's *biophors*, the fundamental vital particles noted above (comparable to de Vries's "pangenes" (as Weismann acknowledged, GP 42) or Hertwig's "idioblasts"), each responsible for determining a particular characteristic of a particular cell. These biophors were themselves capable of assimilation and metabolism, growth, and multiplication by fission (albeit simply in virtue of their physico-chemical constitutions; see GP 37–45), and to account for the diversity of cellular types and characteristics found in the organic world Weismann supposed that they could exist in a nearly unlimited variety of forms produced by differences in the identity and arrangements of their underlying molecular constituents. Despite the fact that they could not be seen under the microscope, Weismann did not regard the existence of such fundamental vital particles as in any way speculative or uncertain:

> *The biophors are not, I believe, by any means mere hypothetical units; they must exist,* for the phenomena of life must be connected with a

> material unit of some sort. But since the primary vital forces—assimilation and growth—do not proceed spontaneously from either atoms or molecules, there must be a unit of a higher order from which these forces are developed, and this can only consist of a group consisting of a combination of dissimilar molecules. I emphasize this particularly, because a theory of heredity requires so many assumptions, which cannot be substantiated that the few fixed points on which we can rely are doubly valuable. (GP 44, original emphasis)

On Weismann's account, then, the continuity of the germ-plasm does not simply consist in a swarm of biophors being reserved from the beginning of ontogeny and then passed from parent to offspring (see esp. GP 64–66). Each biophor in each determinant is a candidate to play a very specific role in the cellular economy of a given type of organism; thus, the reduction division of meiosis does not simply divide the hereditary materials at random, but instead selects a particular *complement* of biophors, hierarchically structured into determinants, ids, and idants, for inclusion in a given sex cell. And while the germ-plasm of any given organism consists of innumerable fragments copied from the "ancestral germ-plasms" of its various predecessors (each such fragment having persisted through the reduction division of each preceding generation of sexual reproduction to be passed down through a reproductive cell in one of the organism's own parents), the mechanics of reduction division ensure that any complete germ-plasm formed by uniting these ancestral fragments to those of another sex cell (as well as the sex cells generated from it in turn) retains a constant proportion of the specific hereditary materials required for the formation of each part of the organism and for each stage of its ontogeny.

But it was not only nor even primarily to ensure the appropriate distribution of these precisely differentiated hereditary materials in reproduction that Weismann attributed a complex structure to the germ-plasm. Weismann embraced Wilhelm Roux's controversial claim (1883; see Coleman 1965 141–142, 152) that as new cells are formed by division in the course of ontogeny the germ-plasm must itself be divided not only quantitatively but also qualitatively. That is, according to what would become known as the Roux/Weismann doctrine, the nuclei of the two cells resulting from cell division in ontogeny typically received *different parts* of the organism's germ-plasm.[12] And Weismann argued that the primary function of the overarching hierarchical structure of any organism's germ-plasm was to ensure that the right components of the hereditary material became available and activated in the right cells at the right time as it underwent these sequential qualitative divisions.

On Weismann's account, then, the fertilized egg's own germ-plasm would be divided in each ontogenetic cell division (each "onto-idic stage"), with qualitatively different portions of the germ-plasm passed on to the various cells destined to generate distinct parts of the body. Thus, an organism's hereditary material was progressively separated into its constituent elements as it directed ontogenetic differentiation and cellular function, and Weismann adopted Nägeli's term "idioplasm" to describe the hereditary substance in this active ontogenetic role (that is, once designated for the control and development of somatic cells and located in their nuclei, with no chance to share in the potential immortality of true germ-plasm). He reserved the term "germ-plasm" (whenever the contrast was relevant) to describe intact, complete *copies* of the hereditary material in the fertilized egg, produced at the beginning of ontogeny and destined to migrate in an inactivated, unalterable state along a specific developmental path into the reproductive cells. This required Weismann to recognize two different (but experimentally indistinguishable) kinds of nuclear division: *homœokinesis*, characteristic of the reduction division required for sex cells, in which the germ-plasm is developed into parts that are structurally identical with respect to their hereditary tendencies, and *heterokinesis*, characteristic of ontogeny, in which the germ-plasm or idioplasm is divided into heterogeneous parts with very different hereditary tendencies. In each ontogenetic cell division into a new onto-idic stage, then, the idioplasm would itself be reduced, although it would not be exhausted by such cell divisions because it would continue to grow throughout the life of the organism (at least while or for any part of the organism in which the cells retained the ability to divide).

Weismann was alone or nearly so in embracing Roux's suggestion that the ontogenetic division of cell nuclei was qualitative as well as quantitative: virtually all of his contemporaries instead followed van Beneden in holding that cell division simply involved a quantitative division and distribution of identical hereditary materials to the two daughter cells. But it would be a mistake to view Weismann as having merely endorsed a suggestion about cell division in ontogeny (and proposed a corresponding structure for the germ-plasm) that failed to pan out. Instead he argued explicitly and repeatedly that there simply was *no possible alternative* to the view that the germinal materials were separated in each ontogenetic cell division until each cell retained only the specific hereditary materials needed to determine its own characteristics. And as the next section will illustrate, it is in Weismann's emphatic insistence on this position in the face of what seemed to his contemporaries to be decisive contrary evidence from cytology and embryology that we begin to see evidence of his failure to conceive of important alternative theoretical possibilities.

5.2 Germinal Specificity, the Search for a Mechanism of Cellular Differentiation, and the Reservation of the Germ-Plasm

Even this brief description of the account of inheritance presented in *The Germ-Plasm* makes clear that Weismann did not share Darwin and Galton's inability to conceive of any alternative to a maturational conception of heredity. Aided by observations of chromosomal behavior unavailable to earlier theorists, Weismann proposed an account of heredity on which germinal materials *controlled* the development and characteristics of the cells they inhabited, but did not themselves *become* the cells, tissues, or other features of the organism whose development they directed. That is to say, Weismann not only conceived of but also endorsed a fully *directive* rather than maturational conception of particulate heredity.

But Weismann's conception of just *how* a particulate hereditary material might exercise this directive function was far from being a simple approximation or less detailed version of our own, a fact reflected in his engagement with the further question of whether the hereditary material is the same or different in the various constituent cells of an organism. He returns repeatedly to this issue in *The Germ-Plasm*, always to insist unequivocally that the nuclei of different cells *must* contain different constituent elements of the organism's hereditary material, a view of the matter that we might call *germinal specificity*.

At first blush, Weismann's insistence on germinal specificity is somewhat puzzling: after all, observations of chromosomal behavior in the nucleus had never suggested any differences between the nuclear materials passed to each daughter cell during ordinary cell division, and every somatic cell appeared to wind up with the same amount of chromatin.[13] But Weismann conceded these empirical facts while brushing them aside as inconsequential:

> It is quite true that the idioplasm of such cells appears similar, at least we can recognize no definable differences in the chromatin rods of two cells in the same animal. But this is no argument against the assumption of an internal difference. The perfect external resemblance between two eggs is not a sufficient reason why two identical chickens should be hatched from them. We cannot perceive these slight differences in either case, and we could not even do so by attempting to analyse the idioplasm concealed in the nuclei of the two eggs by the aid of our most powerful objectives. Theoretical considerations will show later on that it must be so. . . . We shall consequently in this connection have to assume two kinds of nuclear division which are externally indistinguishable from one another, in one of which the two daughter-nuclei contain similar idioplasm, while in the other they contain different kinds of idioplasm. (GP 33–34)

Weismann was surely right to suggest that the failure to detect any difference under the microscope between the nuclei of somatic cells in the same organism did not *rule out* the possibility that such differences existed nonetheless. But how could he be so sure that there *must be* such a difference? What "theoretical considerations" produced such confidence in the absence of any direct supporting observational or experimental evidence?

The primary consideration to which Weismann appeals in arguing for the necessity of germinal specificity is the simple fact that different cells develop differently over the course of ontogeny and ultimately come to exhibit different characteristics:

> The question now arises as to whether all these fragments of the hereditary substance. . . . are similar to, or different from, one another, and it can easily be shown that the latter must be the case. . . . As the thousands of cells which constitute an organism possess very different properties, *the chromatin* which controls them *cannot be uniform; it must be different in each kind of cell.* The chromatin, moreover, cannot *become* different in the cells of the fully formed organism; the differences in the chromatin controlling the cells must begin with the development of the egg-cell, and must increase as development proceeds; for otherwise the different products of the division of the ovum could not give rise to entirely different hereditary tendencies. This is, however, the case. Even the two first daughter-cells which result from the division of the egg-cell give rise in many animals to totally different parts. . . . The conclusion is inevitable that the chromatin determining these hereditary tendencies is different in the daughter-cells. (GP 31–32, original emphasis; see also GP 61)

Likewise, Weismann later argues that the very possibility of ontogenetic development and cellular differentiation depends on the capacity for changes in the controlling idioplasm:

> The idioplasm . . . is capable of regular change during growth; and ontogeny, or the development of the individual in multicellular organisms, depends upon this fact. The two first embryonic cells of an animal arise from the division of the ovum, and continually give rise to differently constituted cells during the course of embryogeny. The diversity of these cells must, as I have shown, depend on changes in the nuclear substance. (GP 45)

Thus, it is first and foremost the simple fact that in the course of an organism's development its various cells come to have very different forms and characteristics that Weismann takes to require germinal specificity.[14] But this seems simply to push our original question back one step. That is, we must now ask how Weismann could be so sure that cells must contain different parts of the hereditary material if their respective courses of ontogenetic development are to diverge.

An important clue to answering this question can be found in Weismann's insistence that it must be by means of the gradual disintegration of the germ-plasm that different cells come to contain different germinal materials over the course of ontogeny:

> In my opinion, it is also an irrefutable fact that this germ-plasm undergoes regular changes from the ovum onwards: it must, indeed, undergo change from cell to cell, for we know that the individual cell is

> the seat of the forces which give rise collectively to the functions of the whole. The forces, which are virtually contained in the germ-plasm can therefore only become apparent when its substance undergoes disintegration, and its component parts, the determinants, become rearranged. The difference in function seen in the various groups of cells in the body compels us to suppose that these contain a substance which acts in various ways. *The cells are therefore centres of force of different worth, and the substance (idioplasm), which controls them must be just as dissimilar as are the forces developed by them.* (GP 204, original emphasis)

Elsewhere Weismann provides a more complete description of this process of the disintegration of the germ-plasm and how it ensures that each cell is provided with precisely the germinal elements needed for its own development. Central to this proposal is the idea that cell divisions separate the idioplasm into simpler and more basic constituent elements:

> As the greater number of these divisions is connected with a diminution in the number of kinds of determinants, the geometrical figure representing the id gradually becomes simpler and simpler, until finally it assumes the simplest conceivable form, and then each cell will contain the single kind of determinant which controls it. The disintegration of the germ-plasm is a wonderfully complicated process; it is a true 'development,' in which the idic stages necessarily follow one another in a regular order, and thus the thousands and hundreds of thousands of hereditary parts are gradually formed, each in its right place, and each provided with the proper determinants.
>
> The construction of the whole body, as well as its differentiation into parts, its segmentation, and the formation of its organs, and even the size of these organs—determined by the number of cells composing them—depends on this complicated disintegration of the determinants in the id of germ-plasm. *The transmission of characters of the most general kind—that is to say, those which determine the structure of an animal as well as those characterizing the class, order, family, and genus to which it belongs—are due exclusively to this process.* (GP 68–69, original emphasis)

Thus it is by the progressive disintegration of the germ-plasm, Weismann argues, that germinal specificity is achieved and that individual cells come to contain the specific elements of the hereditary material appropriate to their functioning and ontogenetic fate.

The significance of this claim becomes apparent once we appreciate the close connection Weismann sees in *The Germ-Plasm* between the question of how the idioplasm is distributed to the respective individual cells of an organism and that of how it achieves and maintains control over each of those cells, for the progressive disintegration of the idioplasm turns out to be the crucial mechanism at work in both of these processes. Concluding the first section of chapter 1, Weismann reminds us that the "capacity on the part of the idioplasm for regular and spontaneous change" is "beyond doubt, when once it is established that the morphoplasm of each cell is controlled, and its character decided, by the idioplasm of the nucleus," and closes with a question: "*But what is the nature of these changes, and how are they brought about*?" (GP 45, original emphasis). The very first sentence of the following section (entitled "The Control of the Cell") assures us that "In order to answer the question which has just been asked, it will be necessary to consider the manner in which the idioplasm of the nucleus determines the characters of the cell" (GP 45).

In this section Weismann takes up and defends de Vries's suggestion that nuclear control of the cell must be mediated by the passage of material particles from the nucleus into the surrounding cytoplasm.[15] And he later goes on to argue that this mechanism for achieving nuclear control of the cell simply *requires* the disintegration of the idioplasm into its constituent elements:

> We have now seen by what means the biophors characteristic of any particular cell reach that cell in the requisite proportion. This results from the fact that the biophors are held together in a determinant which previously existed as such in the germ-plasm, and which was passed on mechanically, owing to its ontogenetic disintegration, to the right part of the body. In order that the determinant may really control the cell, it is necessary that it should *break up into its constituent biophors*. This is an inevitable consequence of the assumed mode of determination of the cell. We must suppose that the determinants gradually break up into biophors when they have reached their destination. This assumption allows, at the same time, an explanation of the otherwise enigmatical circumstance, that the rest of the determinants, which are contained in every id except in the last stages of development, exert no influence on the cell. (GP 69–70, original emphasis)

Here Weismann insists in no uncertain terms that the nuclear idioplasm cannot possibly control the development and differentiation of the cell unless it disintegrates into its constituent material elements. And he maintains this insistence even as he goes on to emphasize our ignorance of the details of the internal structure of the hereditary material itself:

> As each determinant consists of many biophors, it must be considerably larger than a biophor, and is probably therefore unable to pass out through the pores of the nuclear membrane, which we must suppose to be very small and only adapted for the passage of the biophors. Although it is impossible to make any definite statement with regard to the internal structure of the determinants, it must be owing to this structure that each determinant only breaks up into biophors when it reaches the cell to be determined by it. We may suppose that, just as one fruit on a tree ripens more quickly than another, even when the same external influences act on both, so also one sort of determinant may mature

> sooner than another, although similar nourishment is supplied to both.... The assumption of a 'ripening' of the determinants... remains indispensable; or, to express it differently, we must assume that the determinants pass through a strictly regulated period of inactivity, at the close of which the disintegration into biophors sets in. (GP 70)

A few pages later Weismann goes on to insist that the "facts with which we are acquainted" render "unavoidable" the assumption that the germ-plasm can exist in either an "active" or "inactive" state, and simply *defines* the difference between them as consisting in the fact that the former "become disintegrated into their constituent parts" while the latter "remain entire, although they are capable of multiplication" (GP 74–75).

We thus arrive at an answer to our original question about Weismann's confidence in the need for germinal specificity: cellular differentiation over the course of ontogeny absolutely requires germinal specificity by Weismann's lights because he believes that the hereditary material can exert control over the cell only by disintegrating into its constituent elements, and further, that such disintegration can produce cells with different characteristics only if the constituent elements making up the hereditary material in those cells are themselves distinct. That is, Weismann believes not only that the progressive disintegration of the germ-plasm into diverse constituent elements is in fact the process by which germinal specificity is achieved, but also that this is the only possible mechanism by which the germ-plasm *could* control the cell from within the nucleus to produce the kind of cellular differentiation actually observed over the course of ontogeny.

It would be a mistake, however, to suppose that Weismann simply never managed to conceive of any alternative to germinal specificity itself, for the possibility that the *entire* complement of germinal material is duplicated and passed on to each cell of the body in cell division was in fact the view of such important contemporaries as de Vries and Hans Driesch. And indeed, Weismann himself gives clear and elegant expression to this alternative in the course of rejecting it:

> The regularity with which all organs are formed in the proper position and mutual relation, may perhaps be taken as a proof of the assumption that they contain latent determinants which are from the first separate, and which differ according to the topographical position of the organ. It is hardly possible that the contrary assumption can be the correct one, for this would render it necessary to suppose that although all the determinants are certainly present in every formative cell, only that one can undergo development which corresponds to the region in which the cell happened to be situated. (GP 150)

Having so clearly conceived of the idea that the entire germ-plasm is reproduced at each cell-division and contained in the nucleus of each somatic cell of an organism, how could Weismann so confidently dismiss this possibility out of hand? As his more detailed criticisms of de Vries and Driesch make clear, it was because he found it absolutely impossible to conceive of any effective mechanism of ontogenetic development and differentiation that could permit the same hereditary material to reside in the nucleus of each somatic cell.[16]

As we noted above, Weismann endorses a number of de Vries's central claims about heredity from *Intracellular Pangenesis*, including most importantly the proposal that nuclear control of the cell must be mediated by the passage of material particles from the nucleus to the surrounding cytoplasm (GP 46–47, 69). But he insists that it is a profound mistake for de Vries to deny germinal specificity, as this would undermine the very possibility of explaining the ontogenetic differentiation of cells:

> De Vries, on the other hand, considers that the whole of the primary constituents of the species are contained in the idioplasm of every, or nearly every, cell of the organism. But he does not explain how it is that each cell nevertheless possesses a specific histological character. A new assumption, which would not be easy to formulate, would therefore be required to explain why only a certain very small portion of the total amount of idioplasm—which is similar in all parts of the plant—becomes active in each cell. (GP 223; cf. GP 69)

Weismann was well aware that it was processes like regeneration and reproduction by budding in plants which led de Vries to suppose that the entire germ-plasm must be present in every cell. But he insisted that ontogenetic differentiation nonetheless requires different constituents of the hereditary material to be present in different cells, with additional partial or complete copies of the idioplasm (in an inactivated state) invoked as a special adaptation and made available only to particular cells of an organism as needed to explain the abilities of particular parts of organisms to regenerate or to reproduce asexually:[17]

> [My] theory explains the differentiation of the body as being due to the disintegration of the determinants accumulated in the germ-plasm, and requires a special assumption—viz., that of the addition of accessory idioplasm when necessary—in order to account for the formation of germ-cells, and the processes of gemmation and regeneration. The reconstruction of entire plants or of parts from any point can be easily accounted for by de Vries's hypothesis, just as it can by Darwin's theory of pangenesis, for the pangenes or gemmules are present wherever they are wanted. But de Vries is unable, on the basis of his hypothesis, to offer even an attempt at an explanation of the *diversity* of the cells in kind and of the *differentiation* of the body.
>
> These two assumptions appear to me to be of equal value in explaining the fact that in many of the lower plants each cell, under certain circumstances, can apparently reproduce an entire individual. . . . But . . . as soon as the soma can become variously differentiated . . . any explanation must in the first place account for this differentiation: that is to say,

> the diversity which always exists amongst these cells and groups of cells arising from the ovum must be referred to some definite principle. De Vries's principle is of no use at all in this case, for it only accounts for the fact that entire plants may, under certain circumstances, arise from individual cells, and does not even touch the main point. In fact, no one could even look upon it as giving a partial solution of the problem, if differentiation is supposed to be due to that part alone of the germ-plasm always becoming active, which is required for the production of the cell or organ under consideration.[18] (GP 223–224)

Why is Weismann so confident that no explanation of differentiation will be forthcoming on de Vries's assumption that the entire idioplasm is present in the nucleus of each somatic cell? Why, that is, is he so sure that a "further assumption" capable of explaining cellular differentiation without germinal specificity or the disintegration of the germ-plasm "would not be easy to formulate"? To answer this question we will have to consider Weismann's response to an influential series of experiments performed in 1891 by Hans Driesch with newly fertilized eggs of sea urchins, for it is in the course of this response that Weismann argues most explicitly that *no* conceivable mechanism of ontogenetic differentiation could allow precisely the same hereditary material to be present in the nucleus of each somatic cell.

Weismann discusses Driesch's famous experiments in the context of defending the "self-differentiation" of cells; that is, the view that cellular differentiation and development are controlled purely from within the cell and do not occur in response to extracellular stimuli. And he acknowledges that the sea urchin experiments seem to present a challenge for this view. In these experiments, Driesch mechanically separated the cells arising from the first divisions of the fertilized egg and found that the resulting single cells were capable of developing into complete (though unusually small) embryos. And Weismann notes that Driesch takes his experiments to "fundamentally disprove the existence of special regions in the germ which give rise to special organs" (Driesch 1891; quoted in GP 137). In response, Weismann first makes the following somewhat startling claim: "It seems to me that careful conclusions, drawn from the general facts of heredity, are far more reliable in this case than are the results of experiments, which, though extremely valuable and worthy of careful consideration, are never perfectly definite and unquestionable" (GP 138).[19] And the "careful conclusions" to which we must give greater weight than the results of experiments in this case are simply the demands that differential development of the various parts of the organism seem to make for germinal specificity:

> If, however, determinants are contained in the germ-plasm, these can only take part in controlling the formation of the body if, in the course of embryogeny, they reach those particular cells which they have to control,—that is to say, if the differentiation of a cell depends primarily

> *on itself*, and not on any external factor. . . . We can only thereby arrive at the very simple assumptions, that the primary constituents of the germ-plasm are distributed by means of the processes which can actually be observed in the nuclear divisions, so that they come to be situated in those regions which correspond to the various parts of the body, and that those primary constituents are present in each cell which correspond to the parts arising from it. (GP 138)

And although Weismann immediately goes on to acknowledge that reproducing the entire idioplasm in every somatic cell would allow the appropriate germinal material to be available wherever it were needed, he insists that this suggestion is disallowed because it simply forecloses the possibility of *any* conceivable explanation or mechanism of ontogenetic development and differentiation:

> As has just been shown, it is also possible to make the reverse hypothesis, and to suppose that although the whole of the idioplasm is contained in each cell, only that particular primary constituent which properly concerns the cell has any effect upon it. The activity of a primary constituent would thus depend not on the idioplasm of the cell, but on the influences arising from all the cells of the organism as a whole. We should thus have to suppose that each region of the body is controlled by all the other regions, and should therefore practically be brought back to Spencer's conception of the organism as a complex crystal. This simply means giving up the attempt to explain the problem at all, *for we cannot form any conception of such a controlling influence exerted by the whole on the millions of different parts of which it consists*, nor can we bring forward any analogy to support such a view, the acceptance of which would render a great number of observations on the phenomena of heredity totally incomprehensible. (GP 139, my emphasis)

Here Weismann explicitly considers the possibility that the entire germ-plasm is present in the nucleus of every somatic cell and unconditionally rejects it because he cannot conceive of—and indeed judges it impossible to conceive of—any mechanism of ontogenetic differentiation and cellular control that would be consistent with this prospect.[20]

Thus, Weismann's confidence in and insistence on the need for germinal specificity in the face of both the opposing views of his contemporaries and the available experimental evidence rested on a number of distinct failures to conceive of relevant alternative theoretical approaches to particulate heredity. For one, here and throughout Weismann consistently treats the disintegration of the germ-plasm into its diverse constituent elements as the only possible way in which cellular differentiation and ontogenetic development could be directed exclusively from inside the cell. But perhaps even more importantly, Weismann supposes that the only potential alternative to such an internal, disintegrative mechanism of cellular differentiation is the possibility (of which "we cannot form any conception") that the development of cells is controlled by influences coming from *every* other part of the organism. In this judgment he was surely influenced by the fact that his opponents embraced views that appeared to have just this character: Driesch, for instance, writes that "The prospective significance of each blastomere is a function of its position in the whole" (1894 10; cited and translated in Robinson 1979 182) and Hertwig that "all the parts develop in connection with each other, the development of each part always being dependent upon the development of the whole" (1896 105–106; cited in Robinson 1979 182).

Nonetheless, this leaves numerous theoretical possibilities unconsidered, including the one that would prove to be most significant of all for the course of further inquiry: the possibility that the development of various cells containing identical hereditary materials might be differentially affected simply by the varying cues present in their *local* cellular or extra-cellular environments. That is, Weismann seems to have simply failed to consider the possibility (seemingly obvious in retrospect) that the hereditary material is duplicated and passed on intact to each cell in ontogeny and growth, but itself contains or consists of a complex machinery for regulating its own activity in response to different surrounding biological and biochemical conditions. On such a view, different cells could develop quite differently not because different components of the original germ-plasm are present in them, but because different aspects or elements of the identical, complete copies of the original germ-plasm contained in their nuclei are *activated in* or *engaged by* different extra-nuclear and extra-cellular biological environments. Weismann's insistence that the hereditary material contained in the nucleus *must* be qualitatively different in cells that develop differently forces us to conclude that this alternative possibility simply never occurred to him.[21] Instead, as Churchill notes, "that Weismann failed to see clearly a fourth option, namely a morphological totipotency of all cells and a physiological feedback mechanism of activation, suggests the limitations imposed on him by the morphological generalities of the age..." (1987 354n).[22]

Although he alone followed Roux in insisting on a qualitative nuclear division and germinal specificity (and these aspects of his account were widely criticized by his contemporaries) Weismann was anything but alone among theorists of the late- nineteenth century in failing to conceive of any alternative mechanism of ontogenetic differentiation and cellular control. As Coleman remarks, "Only nuclear complexity seemed able to account for growth and differentiation, but how it did so was absolutely unknown" (1965 147; see also Bowler 1989 84–5). Likewise, Dunn suggests that the problems of hereditary transmission could be solved only "when some biologists were willing to put aside the intractable problem of development" and that a convincing account of cellular differentiation would remain elusive: the theory of the gene would be accepted after the turn of the century, he argues, "in spite of the paradox that the mechanism proposed assumed the same variety of units in all cells although the cells themselves became different" rather than because the paradox had somehow been solved or any convincing mechanism of cellular differentiation with identical nuclear material in each cell had been identified (1965 47–48). In fact, the controversy over this aspect of Weismann's view provides us with at least some evidence in favor of the claim that the alternative he failed to recognize was quite *generally* unconceived, for it is surely reasonable to suppose that one of Weismann's many critics on this score (whether opponents of germinal specificity or proponents of general cellular totipotency) would have been delighted to point out this alternative possibility, *if only such a critic had managed to conceive of it himself*. Thus our evidence suggests that neither Weismann *nor those contemporaries who made up his scientific community* ever conceived of or considered this alternative mechanism of inheritance and ontogeny, despite the fact that it was at least as well confirmed by the available empirical evidence as the thoroughly speculative alternative Weismann himself felt compelled to embrace, and sufficiently serious as to have been accepted by later scientific communities (including our own).

It is worth noting that it is because Weismann cannot conceive of any alternative to the disintegration of the germ-plasm into its constituent elements as the mechanism of ontogeny that he is forced to insist on the *reservation* of copies of an individual organism's germ-plasm for its own germinal cells from the very beginning of its development. That is, because the organism's own idioplasm must be disintegrated over the course of its development and in the process of cellular control, Weismann finds himself forced to account for such phenomena as reproduction by budding (GP chap. IV) and the formation of germ cells (GP chap. VI) by assuming that complete copies of the germ-plasm are produced and reserved from the very beginning of its development for this purpose:

> I assume that germ-cells can only be formed in those parts of the body in which germ-plasm is present, and that the latter is derived directly, without undergoing any change, from that which existed in the parental germ-cell. Hence, according to my view, a portion of the germ-plasm contained in the nucleus of the egg-cell must remain unchanged during each ontogeny, and be supplied, as such, to certain series of cells in the developing body. (GP 184)

This "blastogenic idioplasm" consists of one or more complete copies of the organism's germ-plasm, preserved in a special "inactive" and "unalterable" state and passed through particular lineages of cells (the "germ tracks") in the organism's body, ultimately to be located only in its sex cells (after reduction division) and any other cells in a given organism from which offspring may be generated.

Weismann himself appreciates the close connection between this conviction that a special complete copy of the germ-plasm must be reserved for the reproductive cells from the beginning of ontogeny and his own earlier insistence that it must be by means of disintegration of the germ-plasm that ontogenetic development and differentiation is achieved:

> All these facts support the assumption that somatic idioplasm is never transformed into germ-plasm, and this conclusion forms the basis of the theory of the composition of the germ-plasm as propounded here. It is obvious that its composition out of determinants, which gradually split up into smaller and smaller groups in the course of ontogeny, cannot be brought into agreement with the conception of the re-transformation of somatic idioplasm into germ-plasm. If, as we have assumed, each cell in the body only contains one determinant, the germ-plasm—which is composed of hundreds of thousands of determinants—could only be produced from somatic idioplasm if cells containing all the different kinds of determinants which are present in the body were to become fused together into *one* cell, their contained idioplasm likewise combining to form *one* nucleus. And, strictly speaking, even this assumption would be by no means sufficient, for it does not account for the architecture of the germ-plasm: the material only would be provided. Such a complex structure can obviously only arise historically. (GP 190–191, original emphasis)

As we have seen, it is because he can conceive of no alternative mechanism of ontogenetic differentiation and/or cellular control that Weismann is forced to insist that the germ-plasm must disintegrate into its constituent elements in the course of its development. And because he judges it impossible that the organism's germ-plasm could be *re-formed* once disintegrated in this way,[23] this in turn leads him to insist that complete copies of the organism's entire germ-plasm must be reserved for and passed along to its reproductive cells from the very beginning of its ontogeny.

5.3 Invariance, Multiplication, and the Fate of Active Germ-Plasm

As this exploration of both the imaginative failures underlying Weismann's insistence on germinal specificity and some of their consequences makes clear, there is no simple answer to the question whether Weismann shared Galton's inability to conceive of any alternative to an *invariant* conception of particulate heredity or not. On the one hand, Weismann certainly does not share Galton's insistence that the development or activation of any particular hereditary germ must invariably produce a particular characteristic in the organism, nor that a copy of a given germ must invariably generate the same trait in a descendent that it did in an ancestor if it becomes developed or active at all. Instead, although on Weismann's account each id individually contains all the determinants required for the construction and development of a complete organism, any given trait of a particular organism is the outcome of a complex process of competition and interaction among the various constituent elements contained in the many distinct ids making up that organism's idioplasm.[24] Such idioplasm typically includes a large number of homologous determinants (those whose "function is to control the same part of the body," GP 265), each of which may be homodynamous ("impressing a *like character* on any part of the body"; GP 278, original emphasis) or heterodynamous ("tend to impress a somewhat different character on the same part of the body," GP 265) to one another, and which may also vary in their respective degrees of "controlling force."[25] By appealing to a variety of processes of control and competition (including their recombination in sexual reproduction) among such homologous determinants, Weismann is able to offer elegant explanations of any number of observed patterns of individual variation, reversion, the degeneration of characters, the characteristics of interspecific hybrids, changes in the characteristics of a species over the course of its phylogeny, and much else besides. These explanations clearly countenance the possibility that one inherited determinant might interact with, interfere with, or otherwise influence the action of another in such a way as to produce variation at the organismic level: "The power of homodynamous determinants is simply cumulative, whereas dissimilar or heterodynamous determinants may, in the most favourable cases, co-operate to form a single resultant, but may, under certain circumstances, counteract or even neutralise one another" (GP 278). Thus, Weismann clearly avoids Galton's presumption that the end result of the development or activation of any inherited germ cannot depend on what other germs are also inherited or become active.

On the other hand, we have already seen that Weismann does not seem to recognize the possibility that the activity of the germ-plasm (or some particular part thereof) might itself be truly *facultative* or systematically responsive to a range of environmental conditions. That is, while Weismann certainly recognizes that features of the environment can influence what characteristics an organism or cell ultimately comes to exhibit (i.e., GP 107), he does not seem to conceive of the possibility that this might be because the activity of the fully developed germ-plasm *itself* or some group of fully developed hereditary determinants systematically depends on the various cues found in its cellular or extra-cellular environment. This becomes most evident when Weismann considers various ways in which the developmental response of an organism or a constituent cell to its environment *must* itself be facultative, for he can allow for such a response only by multiplying the number of physically distinct idioplasms that are potentially available to become activated and guide the development of the organism or cell in question.

In discussing regeneration, for example, Weismann finds himself forced to assume not only that a cell or type of tissue capable of initiating the regeneration of any parts of the organism distinct from itself must contain a special "accessory idioplasm" ("consisting of the determinants of the parts which can be regenerated by it" (GP 103)) as a dedicated adaptation for this purpose, but also that an organism's cells must contain *multiple* distinct accessory idioplasms of this sort if they are to be able to regenerate in multiple directions (GP 126–127): he notes that in some segmented worms (such as *Nais* and *Lumbriculus*) an amputated part will not only be replaced in the original organism but will also itself regenerate a complete copy of that original organism, and concludes that every cell capable of such bi-directional regeneration must contain two distinct complements of such accessory idioplasm, each of which is supplied with all and only the supplementary determinants needed to produce the rest of the organism in just one direction or the other. As fresh-water polyps and sea anemones are able to successfully regenerate complete organisms from each part of a longitudinal as well as a transverse section, Weismann concludes that the relevant cells of these organisms must each contain *three* distinct accessory idioplasms (one for each spatial direction) again consisting of quite different collections of supplementary ids. In each case, Weismann supposes, the development or activation of just one of these accessory idioplasms is triggered by a "loss of substance" in the appropriate direction.[26] Thus, Weismann can provide for a facultative response by the cell to its environment *only by multiplying the number of different collections of supplementary determinants that might come to control the cell and/or its development*, and not by allowing the response of any given portion of activated idioplasm or of a given collection of developed determinants *itself* to be facultative.

Weismann finds himself similarly forced to multiply the idioplasms that can become activated and take control of the development of a cell in order to account for the various kinds of dimorphism and polymorphism exhibited by organisms. He suggests, for instance, that this must be the case in "dichogeny . . . the form of dimorphism which becomes manifest when a young vegetable tissue, under normal conditions, is capable of developing in different ways according to the external influences to which it is exposed" (e.g. its exposure to light), despite his frank admission that "I can, however, form no idea as to why such an arrangement is met with in this case" (GP 380–382; see also 111, 114). Organisms experiencing alternation of generations must have "*two kinds of germ-plasm* . . . both of which are present in the egg-cell as well as in the bud, though only one of them is active at a time and controls ontogeny, while the other remains inactive" (GP 457, original emphasis; see also chap. V). And sexual dimorphism also "must be due to the presence in the idioplasm of *double determinants* for all those cells, groups of cells, and entire organisms which are capable of taking on a male and female form.... One of the determinants then becomes active, its twin half remaining in an inactive condition in the nucleus of a somatic cell, and under certain circumstances becoming active subsequently" (GP 460–461, original emphasis; see also chap. XI). But such doubling of determinants by no means applies only to secondary sexual characteristics: i.e., because sex-linked diseases like hemophilia occur only in members of one sex, Weismann concludes that the cells of the walls of the blood vessels must also have double determinants, with only the "male" or "female" determinants becoming active in any given individual. And he takes this in turn to be evidence that "*all, or nearly all, the determinants in the human germ are double*, half being 'male' and half 'female,' so that a determinant for any particular part may cause the development of the male or female type of the corresponding character" (GP 372, original emphasis). He accounts for seasonal dimorphism in a single organism in a parallel fashion, while degrees of polymorphism greater than two require further multiplication of the determinants governing each cell, with Weismann ultimately forced to assume that some kinds of bees and termites have triple and even quadruple determinants in their cells, only one set of which becomes developed or activated in any given individual.[27] Weismann recognizes that he is thus forced to posit an "ever increasing complexity of the substance which renders repetition of the organism possible," but insists that "it is impossible to explain the observed phenomena by means of much simpler assumptions" (GP 468). He seems, that is, to recognize no alternative to *encapsulating* an organism's or cell's developmental response to a particular set of circumstances in a physically distinct accessory germ-plasm that itself simply takes over and becomes the controlling idioplasm of the cell under the appropriate conditions, distintegrating into its constituent elements as it guides development and differentiation.

This inability to conceive of an idioplasm capable of a facultative response to its environment not only forces Weismann to multiply the physically distinct idioplasms which might come to control a cell under various kinds of circumstances, but also forces him to insist that any substantial change in the functioning or operation of a cell must be accompanied by a corresponding change in the makeup of its controlling idioplasm. This feature of Weismann's account becomes especially salient in the course of his discussion of the expulsion of the polar bodies from the egg cell during oogenesis.[28] There he first argues that the formation and histological development of the egg cell must be governed by a special kind of dedicated "oogenetic idioplasm:"

> If the nature of the cell is determined at all by its idioplasm, the ovum, while still growing and undergoing histological development, cannot possibly be controlled by the same idioplasm as that which serves for embryonic development. I consequently assumed the existence of an '*oogenetic*' idioplasm in the egg during the period of its histological differentiation, and also that after maturation this substance gives up control of the cell to the germ-plasm.
>
> The question then arises as to what becomes of the oogenetic idioplasm when this change in the control takes place. (GP 349, original emphasis)

Weismann's own earlier answer to this question had been that the oogenetic idioplasm must be *expelled* from the egg-cell to prevent it from interfering with the development of the fertilized egg and that this removal of the oogenetic idioplasm was itself the function of the expulsion of at least one of the polar bodies from the egg during oogenesis. He here allows that new evidence has shown this view of the matter to be mistaken and that the expulsion of the polar bodies does not involve the removal of a special oogenetic idioplasm.[29] But this does not lead him to question the existence of a special oogenetic idioplasm in the first place, and in fact the recognition that no such idioplasm is expelled from the maturing egg cell simply allows Weismann to say with certainty what the fate of this oogenetic idioplasm must be. He continues this section, entitled "Proof that the Determinants become Disintegrated into Biophors," by concluding that the oogenetic idioplasm must instead be *consumed* in the course of performing its directive function:

> The oogenetic idioplasm must exist, and, using the terminology I have now adopted, it may be spoken of as the oogenetic 'determinant.' This determinant will consequently be the first to become separated from the mass of germ-plasm of the young egg-cell, to disintegrate into its constituent biophors, and to migrate through the nuclear membrane into the cell-body. In this way alone can we account for no trace of it remaining in the nucleus, and for embryonic development not being subsequently impeded by its presence. *This determinant is used up, and disappears as such*; and the fact that it is not expelled from the egg strongly indicates, if it does not prove, that the control of a cell by a determinant is accompanied by the absorption of the latter.... (GP 350, original emphasis)

Most notable here, of course, is Weismann's insistence that whatever germ-plasm is responsible for the development of the egg cell could not continue to exist in the egg without interfering with the formation of the developing embryo. That is, *given* that the oogenetic idioplasm is not expelled, it *must* instead be used up in the course of the formation and development of the egg itself and therefore not remain in the egg when the functioning of the latter radically changes. And Weismann confidently extends the lesson learned in this particular case to the operation of the germ-plasm in general:

> I know of no instance in which there is such a wide difference as regards the activity of the idioplasm in successive cell-generations as is the case in the germ-mother-cells and the mature germ-cells arising from them. If, however, even in this very striking instance of a sudden change of function of the idioplasm, the idioplasm which was active at first is not removed from the cell, such a process cannot occur in any other case; and we are consequently justified in applying to all other cells the conclusion derived from the behavior of the germ-cells, and in considering it as proved that the *active idioplasm of a cell becomes used up in consequence of its activity.* (GP 351, original emphasis)

The most important point here is Weismann's presumption that the germ-plasm or any given portion thereof, once developed or activated, is forced to continuously exert a particular effect on the cell in which it resides until physically expelled, destroyed, or exhausted. Thus, Weismann's failure to conceive of even the possibility that the germ-plasm might be capable of systematically regulating its own activity in response to the conditions present in its cellular or organismic environment forces him not only to multiply the physically distinct idioplasms that may come to control a cell in order to allow for any facultative response of a cell to its environment, but also to argue that once activated the idioplasm of a cell must be used up (since it is not expelled) in the course of exercising its directive function. And these features of Weismann's account and the arguments he makes for them illustrate the important respects in which he himself remains unable to conceive of any alternative to an invariant conception of heredity, despite the clear progress he was able to make over Galton's imaginative imprisonment by this conception.

5.4 Productive and Expendable Germinal Resources

In the final analysis, however, it seems natural to suggest that the two central failures of theoretical imagination we have seen in Weismann, as well as the further consequences we have noted for his account of inheritance and generation, are themselves rooted in a further and still more fundamental inability to conceive of alternative theoretical possibilities. More specifically, *both* Weismann's failure to conceive of any alternative to the disintegration of the idioplasm as the mechanism of ontogenetic differentiation and nuclear control *and* his failure to conceive of any genuinely facultative capacity on the part of the germ-plasm suggest in turn that Weismann never conceived of the quite general possibility that the germ-plasm could itself serve as what we might call a *productive* rather than an *expendable* resource for the cell and/or the organism.

That is, Weismann seems to conceive of the germ-plasm as itself necessarily *consisting of* a bundle of material resources to be used in controlling the development and differentiation of cells, and he seems never to consider the possibility that the germ-plasm might instead represent the cell's (or the organism's) own machinery for *generating* or *producing* such materials.[30] Consider, for example, the further inference Weismann draws from establishing to his own satisfaction that nuclear control of the cell must be mediated by the passage of material particles from the nucleus into the surrounding cytoplasm of the cell:

> If then, each vital unit in all organisms, from the lowest to the highest grade, can only arise by division from another like itself, an answer is given to the question with which we started; and we see that the structures of a cell-body, which constitute the specific character of the cell, cannot be produced by the emitted influence of the nuclear substance, nor by its enzymatic action, but can only arise owing to the migration of material particles of the nucleus into the cell-body. *Hence the nuclear matter must be in a sense a storehouse for the various kinds of biphors, which enter into the cell-body and are destined to transform it.* Thus the development of the 'undifferentiated' embryonic cell into a nerve-, gland-, or muscle-cell, as the case may be, is determined in each case by the presence of the corresponding biophors in the respective nuclei, and in due time these biophors will pass out of the nuclei into the cell-bodies, and transform them.
>
> To me this reasoning is so convincing that any difficulties we meet with in the process of determining the nature of the cell hardly come into account. (GP 48–9, original emphasis)

As this image of a nuclear "storehouse" suggests, Weismann here confidently treats the view that nuclear control of the cell must be mediated by the passage of material particles from the nucleus to the surrounding cytoplasm as tantamount to assuming that the germinal material must *itself consist of* such particles and therefore undergo disintegration and pass out of the nucleus in the course of controlling the cell. He seems never to consider the possibility that the role of the germ-plasm could instead be manufacturing the necessary materials for transmission to the cytoplasm, much less that it could do so in a systematically facultative way. That is, he never considers the possibility that the germ-plasm might represent a sort of biochemical *factory* able to produce materials for controlling the functioning and development of particular cells in response to varying conditions in the local environment, and not the organism's *supply* or *stockpile* (or "storehouse") of such materials.

It would seem, then, that it is ultimately because Weismann is constrained to think of the germ-plasm as an expendable resource for the cell that he cannot conceive of any alternative to its disintegration as the mechanism of ontogenetic differentiation and cellular control; after all, conceiving of it instead as a productive resource quite naturally suggests that the germ-plasm would generate rather than consist of whatever material particles pass into the surrounding cytoplasm in order to mediate these processes. And as we have seen, it is because Weismann cannot conceive of any alternative to the disintegration of the germ-plasm that he is in turn forced to insist on central doctrines such as germinal specificity and the reservation of the germ-plasm from the beginning of ontogeny. In a similar fashion, it would appear to be because he is constrained to conceive of the germ-plasm as an expendable resource that Weismann fails to conceive of the possibility that the hereditary material might be capable of mounting a truly facultative response to its environment. It is because the germ-plasm simply consists of the bundles of material resources it might use to effect differentiation and control that Weismann is forced to regard the activation of one rather than another physically distinct and encapsulated ordered sequence of such resources as the only kind of response to a biological or biochemical environment that a cell or nucleus can exhibit. In the grip of this presumption, as we saw, Weismann is forced to provide for systematic variability in the form and function of a cell only by multiplying the various expendable idioplasms that might ultimately come to control it and to insist that activated or developed germ-plasm, since it is never expelled from the cell, must be consumed in the course of exercising its directive function. By contrast, conceiving of the germ-plasm as a productive resource seems to fairly invite the notion that it acts as a persistent physical intermediary between specific conditions in the extracellular or extranuclear environment and the specific directive material responses provided by the nucleus itself.

5.5 Conclusion: Lessons from History

Weismann's failures to conceive of serious alternative theoretical possibilities illustrate with striking clarity a kind of nested hierarchical structure that we have also seen suggested by the cases of some earlier theorists. That is, it would seem to be at least in large part *because* Weismann fails to imagine that the hereditary material might be a productive rather than expendable resource that he fails in turn to conceive of any possible alternative to disintegration of the germ-plasm as a mechanism of ontogenetic differentiation and cellular control or of the possibility that the germ-plasm itself might be capable of a systematically facultative response to its local environment. And these failures of theoretical imagination lead in turn, as we've seen, to Weismann's insistence that a specially inactivated germ-plasm must be reserved for the reproductive system from the beginning of ontogeny, that physically distinct germ-plasms or idioplasms must be multiplied in cells capable of responding facultatively to their environments, and that the idioplasm must itself be consumed (because it is not expelled) in the course of directing the development and activity of the cell. This nested structure of connected theoretical inferences helps to make clear that the significance of the challenge posed by unconceived alternatives does not ultimately depend (as it might have seemed at first glance) on the blanket claim that the theoretical possibilities we have regarded as neglected were *never* conceived of in *any* way or at *any* time either by a particular scientist or by *any* members of the relevant scientific community. Even if we were to uncover heretofore unknown evidence that Weismann (or Hertwig, or de Vries, or others) did in fact catch a momentary glimpse of some of these neglected possibilities through a glass darkly, they do not seem to have been taken into account *when it mattered in this case*, that is, at the time Weismann was willing to draw and trying to justify significant inferences and conclusions about the nature and constitution of the hereditary material, about the proper course of further research, and about what the processes of inheritance and generation must be like. And it is enough to threaten our eliminative practices of confirmation that Weismann (or Darwin, or Galton, respectively) neither conceived of nor considered the relevant theoretical alternatives when it really counted in this way.

On the other hand, what ultimately matters of course is not whether individual scientists are able to exhaustively consider the space of well-confirmed alternative theoretical possibilities, but whether scientific *communities* are able to do so. As a general matter, the failure of a given individual scientist to conceive of or consider particular theoretical alternatives serves us simply as *evidence* that the relevant alternatives were not conceived of or widely considered in the community at large. This, of course, is why it was especially important to take note of the historical evidence supporting the claim that de Vries, Driesch, Hertwig, and other opponents of germinal specificity failed to conceive of the general type of mechanism of ontogenetic differentiation that eluded Weismann as well. But we should not make the mistake of thinking of even the case of Darwin and Galton as one in which a community of scientists working together were somehow able to exhaust the space of well-confirmed possibilities when one alone was not. While it is true that Galton managed to conceive of one specific possibility that Darwin failed to grasp, that is not to say that between the two of them Darwin and Galton managed to exhaust the space of well confirmed alternative theoretical possibilities, or even just those that would later be embraced by some actual scientific community: even as Galton managed to conceive of the common-cause conception of inheritance that had eluded Darwin, Weismann's directive conception of particulate heredity and much else besides remained entirely unconceived by Darwin, Galton, or any of their peers.

Much traditional history of biology regards the latter decades of the nineteenth century as a period in which the unavoidably speculative excesses of earlier theorists were abandoned and/or replaced by alternatives increasingly grounded in or constrained by the hard empirical facts uncovered by advances in microscopical observation and embryological experimentation. It is well worth noting, then, that Weismann's failures to conceive of important theoretical alternatives that were no less well supported by the available evidence undermines at least one related route of response to the problem of unconceived alternatives itself. It might have seemed reasonable enough to suppose that unconceived alternatives only represent a serious problem for theoretical science when we are unable to directly observe or detect the central objects of our theorizing. As difficult as it has turned out to be to characterize the relevant notions of direct observation or detection rigorously, it might nonetheless have seemed that more-or-less direct observational contact with the entities about which we are theorizing serves to radically constrain the space of serious and well-confirmed theoretical hypotheses in such as way as to eliminate any real danger posed by the problem of unconceived alternatives. But the case of Weismann makes this strategy of response to the problem look unpromising: after all, Weismann was himself among those who knew that they had managed to observe the hereditary material through the microscope and to track its changing character through such crucial processes as cell-division, fertilization, and the formation of gametes, and he made extensive use of the latest observations in nuclear cytology to argue for and against particular claims about the processes of inheritance and generation. Nonetheless, Weismann remained unable to conceive of important theoretical possibilities concerning any number of aspects of this hereditary material, including its constitution, its operation, and its most fundamental character. Thus, even the ability to engage in detailed and systematic observation or detection (in the standard scientific senses of those terms) of the objects of our theorizing seems to offer no proof against the relevance or centrality of the problem of unconceived alternatives.[31]

Nor, it would seem, does the ability to make successful novel predictions in a given domain of theorizing indicate that we are beyond the reach of the problem of unconceived alternatives, despite the currency of this notion in much recent philosophy of science. Weismann's prediction of the need for reduction division in the formation of the sex cells still stands as one of the classic cases of confirmed theoretical prediction of a previously unknown phenomenon in the history of biology, and it was recognized as such even by his contemporaries (see Robinson 1979 182–183). Nonetheless, Weismann managed to make this surprising novel prediction—about the behavior of a hereditary material that had not yet even been conclusively identified—while failing to conceive of important theoretical alternatives to his own views of the operation, constitution, and fundamental character of that hereditary material itself.

The evidence we have seen makes it similarly unpromising to suggest that the unconceived alternatives neglected in the historical course of our theorizing about generation and inheritance have become somehow progressively less fundamental or significant over time. While it is undoubtedly true that Weismann's account of heredity is closer to our own than was Darwin's, and the historical developments we have considered are often Whiggishly portrayed as a story of incremental progress on matters of increasingly minute detail in the course of a natural evolution toward our own contemporary view of inheritance, actually identifying past scientists' specific failures to conceive of unconceived alternative possibilities belies the suggestion that the alternative possibilities neglected by past theorists themselves represent progressively less and less fundamental divergences from a contemporary view or from the space of theoretical possibilities already under consideration. Darwin's failure to recognize the possibility of a common-cause structure for hereditary resemblance surely represents a fundamental failure to conceive of alternative possibilities, but not obviously more so than Galton's inability to conceive of a contextual rather than invariant conception of heredity. And their shared failure to conceive of a directive rather than maturational conception of heredity seems to neglect alternatives that diverge neither clearly more nor clearly less fundamentally from the remaining possibilities (including our own account) than those excluded from consideration by Weismann's inability to conceive of the hereditary material as a productive rather than expendable resource. In a similar vein, we might note that it seems relatively easy to imagine an historical sequence of discovery in which, for instance, the possibility that the hereditary material is productive was recognized before the possibility that inheritance has a common-cause structure. Thus, while the failures to conceive of serious alternative possibilities exhibited by earlier theorists may often or even characteristically include those of later theorists, this seems to entail *neither* that the range of alternatives excluded by these later failures diverge in less central respects from the space of considered possibilities *nor* that the consequences of neglecting them are any less confirmationally significant. This fact also encourages us to be deeply suspicious of the idea that past theorists simply had no hope of discovering the correct account of inheritance and generation without today's sophisticated molecular chemistry (cf. chap. 3, Sec. 1), and that the development of such a modern chemistry was the breakthrough that finally brought this search to a successful conclusion. For the historical evidence we have considered gives us every reason to wonder what presently unconceived alternatives are playing the same role for today's scientists that contemporary molecular chemistry played for theorists of the past. That is, we have every reason to believe that there are theoretical alternatives remaining unconceived by us whose grasp will be regarded by future scientific communities as absolutely fundamental and/or a necessary precondition for conceiving of or even understanding the further accounts of nature that they themselves embrace.

Our extended historical discussion has also helped to clarify what would be required in order to resist projecting the new induction and its associated problem of unconceived alternatives from past to present and/ or future science. As we noted in chapter 2, we would have to identify relevant differences between earlier *theorists* and ourselves, rather than between earlier theories and our own. That is, to think that past scientists were subject to the problem of unconceived alternatives but that we ourselves are not, we would have to believe that the institutions or practitioners of scientific research have themselves changed in such a way as to somehow permit us to exhaustively consider the space of well-confirmed alternative possibilities. We might think, for instance, that the simple fact that there are many more scientists working today than in the past renders us much more likely to discover any well-confirmed theoretical alternatives that do exist. This strikes me as an interesting but ultimately implausible suggestion, in part because I do not see any reason to suppose that the relatively larger number of scientists at work in the present day should enable us to *exhaust* the space of well-confirmed theoretical alternatives, as opposed to simply hastening the process of discovering previously unconceived ones. But far more important is the fact that the theoretical possibilities that matter here are *fundamental* alternatives to the dominant account of nature embraced at a given time. While the professionalization of contemporary science has indeed ensured that there are many more practitioners at work exploring various aspects of our present accounts of the natural world, I see no reason to think that it has rendered them *collectively* less vulnerable to the kinds of conceptual barriers or limitations that (as we have seen) prevented Darwin, Galton, and Weismann from conceiving of the relevant well-confirmed fundamental alternatives to the theoretical accounts of nature they considered. Indeed, it is far from clear to me that the professionalization of contemporary science has not produced an incentive structure encouraging considerably *less* investigation of fundamentally distinct alternatives to dominant theoretical accounts of nature than when much of our scientific inquiry was conducted by independently wealthy gentleman-scholars like Darwin and Galton pursuing their own intellectual agendas.

There is perhaps a further point worth emphasizing in this connection. Although it is always in light of later theoretical developments that earlier failures to conceive of possible alternatives become easy to see, and we have here confined our attention to unconceived alternatives that would later be embraced by some actual scientific community, we have by no means merely pointed out the various ways in which the space of theoretical possibilities conceived of by Darwin, Galton, and Weismann respectively failed to make room for contemporary molecular genetics: the space of important theoretical possibilities neglected by each of the theorists we have considered included much else besides the views of inheritance that we ourselves embrace. Darwin, for instance, failed to conceive of the possibility of any common-cause structure for inheritance whatsoever: the space of alternative possibilities thereby neglected would include both Galton's and Weismann's fundamentally mistaken accounts of heredity and many other possibilities besides in addition to our own contemporary account. Galton failed to conceive of the possibility of hereditary mechanisms that were directive (rather than maturational) or in any way contextual (rather than invariant), a set of alternatives that would likewise include Weismann's mature view and much else besides our own account of the matter. And Weismann's own inability to conceive of *any* alternative to the disintegration of the germ-plasm as the fundamental mechanism of ontogenetic development and cellular differentiation, of the capacity for a facultative response of the germ-plasm to its local environment, and of the germ-plasm itself as a productive rather than expendable resource leaves aside a much broader range of theoretical alternatives than just the particular accounts that we ourselves have come to adopt. In each case, an indefinitely large *space* of important theoretical alternatives appears to be neglected, albeit one that happened to include views (including contemporary molecular genetics) that would in fact be embraced by some later scientific community.

It is in large part because the space of unconceived alternative theoretical possibilities appears to have this characteristic structure that the problem seems to persist even though we are capable of genuine conceptual improvement on past science and therefore can ourselves enjoy the luxury of conceiving of and considering an ever-larger space of serious theoretical alternatives. Of course, even if the space of unconceived alternatives contained only a finite number of well-defined possibilities, we would seem to have little reason to believe that we are presently at the end of an exhaustive search of it and have finally reached the point at which serious unconceived possibilities no longer pose any real danger to our theoretical science in a given domain. But even aside from this, the space of serious theoretical alternatives would seem to have a vague and indefinite character, with members that are difficult if not impossible to individuate sharply or unequivocally: an indefinite number of alternative possibilities are neglected by failing to consider the possibility that heredity has a common-cause structure, but an indefinite number of serious possibilities appear to remain excluded if we recognize this possibility but fail to consider the further possibility that the hereditary material is a productive rather than expendable resource. And if (or wherever) the space of serious theoretical possibilities in which we seek to apply eliminative tools of confirmation appears to be indeterminate and unbounded in this way, it seems that we can have little confidence in the power of our eliminative inferences to arrive at the theoretical truth of the matter regardless of how (finitely) long we allow them to operate.

This progressive unveiling of an indefinitely contoured and ordered space of unconceived alternatives also helps to illustrate why another plausible early challenge for the new induction has proved less of an obstacle than it promised to be at the outset. We might initially have worried, naturally enough, how supporters of the new induction could be in a position to judge that a given alternative was even roughly as well confirmed as the extant competitors by the evidence available at the time it was unconceived. But attention to the details of the historical examples we have considered have made this concern seem misplaced as well. The respects in which the fundamental lines of theorizing unconceived in our examples have diverged from those accepted or defended at the time have typically been ones concerning which there simply was no available evidence (surely in large part *because* the relevant alternatives were unconceived): for instance, whether hereditary similarity is produced by a common cause rather than a causal chain, whether inheritance and generation are directive rather than maturational or contextual rather than invariant, and whether the hereditary material is a productive rather than an expendable resource. Thus, we need not discharge the difficult task of showing that fundamentally distinct alternatives were at least roughly equally well confirmed by the available evidence according to some specific (and unavoidably contentious) standard of confirmation, because the details of our examples leave us little room to doubt that the relevant serious alternatives were quite genuinely unconceived at the time and not simply dismissed or ignored for lack of evidential support.

Of course, it is in no way *surprising* that Darwin, Galton, Weismann, or any of their contemporaries failed to conceive of the alternative possibilities that eluded them, for scientists neither do nor claim to proceed generally by surveying all *possible* theories before trying to confirm one that has occurred to them against the extant alternatives. Nor are we suggesting that these thinkers were somehow irresponsible or careless, either in failing to conceive of the particular serious theoretical alternatives they neglected or in being willing to draw eliminative inferences from what they saw as the only possible accounts or mechanisms of inheritance and generation that could accommodate the evidence. Such inferences are perfectly legitimate ones in a wide variety of epistemic contexts, and in each case it was difficult theoretical work of the highest scientific order to conceive of the relevant unconceived possibilities. Furthermore, it will represent a significant epistemic discovery if we ultimately conclude that fundamental scientific theorizing is simply not among the contexts in which the conditions necessary for reliable eliminative inferences are satisfied. Thus the moral is not that Darwin, Galton, or Weismann made reckless inferences or that any of them didn't realize something that he should have realized—after all, the suggestion here is that such blindness to serious theoretical alternatives characterizes the activity of fundamental scientific theorizing quite generally, and each of our theorists managed to recognize important theoretical possibilities neglected by earlier thinkers—but rather that human beings are simply not good at conceptually exhausting the space of serious alternative possibilities in the context of fundamental scientific theorizing, and are therefore not entitled to believe the conclusions of their eliminative inferences in this context.

This does not mean the moral suggested here is that we must somehow constrain and regulate the inferences we draw in the course of our scientific theorizing by some perfectly abstract and general commitment to the likely existence of completely unspecified but serious unconceived theoretical alternatives. What could this amount to but either a sure recipe for inferential (and therefore conceptual) paralysis or a vapid agreement to tack the phrase "but of course there may be something I haven't thought of yet" piously and toothlessly onto every conclusion we draw? Darwin, Galton, and Weismann were all quite right to draw substantive conclusions from what they believed to be the only possible mechanisms of generation and inheritance—this is an important way in which much productive scientific theorizing works and no useful replacement for or revision of this aspect of scientific methodology has been proposed or defended here. Instead, the moral is that we must adjust both our sense of what theorizing that relies on such an eliminative inferential methodology is capable of achieving and our attitude toward the epistemic status of the *products* of such theorizing. We have abundant evidence that such eliminative inferences can guide us to theories that are powerful instruments for mediating our engagement with the natural world and that confer upon us powers of prediction, intervention, and putative explanation undreamed-of by earlier generations. But it would seem that we have equally abundant (albeit less direct) evidence that the problem of unconceived alternatives prevents such eliminative inferences from being reliable guides to the *truth* concerning the otherwise inaccessible domains of nature about which we theorize.

Notes

1. Coleman (1965) offers an extremely useful detailed historical discussion of the developments in cytology described in this paragraph; see also Churchill (1968 103f), Robinson (1979 137–141), and Bowler (1989 85–87).

2. Although Schneider's results were not immediately well-known, similar observations were soon made by Bütschli, van Beneden, and Fol (see Coleman 1965 131, Robinson 1979 137).

3. While Hertwig reported that the two nuclei fuse in the process of fertilization, the later work of van Beneden would reveal that the respective nuclear contributions from sperm and egg remain intact and discrete (see Coleman 1965 140–141, Churchill 1968 106).

4. As any number of historians have argued and as the remainder of this chapter will illustrate, however, these dramatic new observations were by no means sufficient to resolve any of the central controversies about inheritance or generation by themselves. Indeed, at this point the chromosomes were conceived

of primarily as morphological rather than physiological entities, and even their physiological continuity between cell divisions was debated well past the turn of the twentieth century. Furthermore, biologists continued to disagree stridently and sometimes vituperatively among themselves about the significance and implications of these cytological findings, and even about what entities and processes had in fact been observed under the microscope. As Bowler notes, it would be "the growing popularity of the explanatory system which became the basis of classical genetics that at last allowed biologists to agree over the interpretation of their observations" (1989 86).

5. It is important not to exaggerate the extent to which this initial formulation of Weismann's account was itself conceived in reaction to or even constrained by the ongoing developments in nuclear cytology: his central concern at the time was instead with problems of evolution and the transmission of characters from parents to offspring (i.e., the inheritance of acquired traits). Indeed, he would later note that when the Freiburg address was written, "I was not aware that this germ-plasm existed only in the nucleus of the egg-cell, and I was therefore able to contrast the entire substance of which the egg-cell consists, or the germ-plasm, with the substance which composes the body-cells, hence called somatoplasm" (1890 83). And it was not until the years between 1883 and 1885 that Weismann, Hertwig, and Strasburger would independently focus their attention on the role of the cell nucleus in heredity (see Coleman 1965 140, Robinson 1979 141).

6. Weismann reports that he came to this belief in the continuity of the germ-plasm under the impression that the theory was entirely original, but later discovered that "similar ideas had arisen, in a more or less distinct form, in other brains" ([1892] 1893 198). Most notably, he grants that Galton's ideas "bore some resemblance to the conception of the continuity of the germ-plasm," but nonetheless insists that there are crucial differences between his own proposal and Galton's, perhaps most importantly that his own version of the idea does not depend upon a "residue" left over after hereditary particles are selected for development, but instead "is founded on the view of the existence of a special adaptation, which is inevitable in the case of multicellular organisms, and which consists in the germ-plasm of the fertilized egg-cell becoming doubled primarily, one of the resulting portions being reserved for the formation of germ-cells" ([1892] 1893 200).

7. Most famously, Weismann had predicted the need for reduction division in the formation of sex cells on the basis of purely theoretical considerations. He would not, however, immediately accept van Beneden's suggestion that this reduction was effected in egg cells by the expulsion of the polar bodies, interpreting the latter event instead as the removal of a special "oogenetic idioplasm" (see Weismann 1885, 1887, and Churchill 1968 106–108).

8. This was perhaps a natural consequence of Weismann's unique position at the intersection of evolutionary theory and cytology (see Bowler 1989 84f), where he combined "the points of view of the microscopical anatomist, the embryologist, and the evolutionist" (Coleman 1965 152).

9. Haeckel argued, that is, that reproduction simply reflected the separation and continued growth of surplus material from the parents' own body or bodies, and thus that "reproduction is a maintenance and a growth of the organism over

and beyond the individual mass, one part of which is elevated to the whole" (Haeckel 1866 II 16; cited in Churchill 1968 97).

10. Weismann also finds it "probable that the ids correspond to the small granules hitherto called 'microsomata,' which are known to form the individual idants in many animals" (GP 67; see also GP 240–241). Note that Weismann (like de Vries) was by this time primarily a theorist rather than a microscopist, as the deterioration of his eyesight had forced him to abandon his own microscopical research many years earlier (see Coleman 1965 151).

11. Though Weismann allowed that even large groups of identical cells might be represented in the germ-plasm by just a single determinant, any two cells capable of independently heritable variation would have to be represented in the germ-plasm by distinct determinants (GP 53–57).

12. Although Roux originally proposed the notion that the quantitative division of nuclear material in cell division was also a qualitative division in 1883, Weismann would not credit Roux with this idea until 1887 (see Churchill 1968 103n).

13. Indeed, Van Beneden's claim (see above) that the nuclear division was symmetric and quantitative was influential precisely because of the persuasiveness of his experimental work. Other theorists (e.g. Kölliker and Strasburger; see Robinson 1979 151–4 and 159–60) would likewise argue that Weismann's insistence on germinal specificity and on an important difference between the germ-plasm contained in various cells (e.g. germ cells and soma) was *cytologically* implausible.

14. This argument for germinal specificity appears in Weismann's work at least as early as 1885: "I therefore believe that we must accept the hypothesis that, in indirect nuclear division, the formation of unequal halves may take place quite as readily as the formation of equal halves, and that the equality or inequality of subsequently produced daughter-cells must depend on that of the nuclei" (1885 193).

15. More fully, Weismann argues that if the idioplasm is to "exert a determining influence" over the cell, "it must either be capable of exerting an emitted influence (*Fernwirkung*) or else material particles must pass out of the nucleus into the cell body" (GP 45). But he argues that the first possibility would require the structures of living cells to come into existence by "a kind of *generatio equivoca*" in which "they would have arisen by the operation of an external influence on the given substance in the cell, just as would be the case in primordial generation." And Weismann insists not only that such primordial generation is unknown in "those forms of life with which we are acquainted," which "always arise by division from others similar to themselves," but also that "[w]e can only imagine the very simplest biophors as having been produced by primordial generation: *all subsequent and more complex kinds of biophors can only have arisen on the principle of adaptation to new conditions of life*" (GP 47–48, original emphasis). Thus "the structures of a cell-body, which constitute the specific character of the cell, cannot be produced by the emitted influence of the nuclear substance, nor by its enzymatic action, but can only arise owing to the migration of material particles of the nucleus into the cell-body" (GP 48–49, original emphasis).

16. Elsewhere Weismann dismisses this possibility in a more offhanded way, suggesting that it violates the principle that "Nature . . . always manages

with economy" (GP 63), but his detailed argument that ontogenetic differentiation requires different cells to contain different hereditary materials is a considerably more developed and fundamental part of his thinking. Furthermore, we will see that Weismann is ultimately forced to drastically multiply unused and inactive partial or complete copies of the germ-plasm to accommodate a wide variety of forms of facultative responsiveness that cells exhibit to their environments, which makes his selective appeal here to the economy of nature look suspiciously opportunistic.

17. And Weismann repeatedly emphasizes that regeneration or budding can only be initiated from some cells of an organism and not others—a fact that he regards as further evidence for germinal specificity.

18. In the secondary literature, differences between Weismann and de Vries have often been ascribed to the fact that Weismann was a zoologist while de Vries was a botanist and each was most impressed and concerned with the hereditary phenomena characteristic of the types of organisms that he had studied most closely (see e.g. Robinson 1979 175, Bowler 1989 91). It is therefore interesting to see Weismann here explicitly describing his own account as *equally* able to explain characteristic botanical phenomena like budding, as well as better able to explain the phenomena of cellular differentiation, and going on to suggest that de Vries is but he himself is not a victim of the professional provincialism implicit in this contrast: "But the higher we ascend in the organic world, the more limited does the power of producing the whole from separate cells become, and the more do the numerous and varied differentiations of the soma claim our attention and require an explanation in the first instance.... In the lower plants the fact of the differentiation of the soma is liable to be overlooked or underrated, but this cannot possibly be the case as regards the higher animals" (GP 224). Interestingly, de Vries makes a similar accusation in the reverse direction regarding modes of reproduction in *Intracellular Pangenesis* ([1889] 1910 81).

19. Perhaps this methodological injunction will seem less surprising if we recall that Weismann had been forced by the progressive deterioration of his eyesight to abandon his own microscopical research many years earlier (see above and Coleman 1965 151).

20. Of course, neither Driesch nor de Vries (nor any other theorist of this period) had actually *proposed* an alternative mechanism of cellular differentiation of the sort whose very possibility Weismann failed to conceive. We will return to this point and its significance later in the chapter.

21. Moreover, Weismann's failure to recognize this possibility is evident at least as early as the 1883 Freiburg address, in which he wrote (while still in the grip of Haeckel's "overgrowth" conception of reproduction; see Churchill 1968): "as their development shows, a marked antithesis exists between the substance of the undying reproductive cells and that of the perishable body-cells. We cannot explain this fact except by the supposition that each reproductive cell potentially contains two kinds of substance, which at a variable time after the commencement of embryonic development, separate from one another, and finally produce two sharply contrasted groups of cells" (1883 74).

22. Nonetheless, Churchill also rightly points out that Weismann's justification for his conclusions was explicitly eliminative in character: "It is difficult to determine from the printed sources which of Weismann's many arguments for

continuity [of the germ-plasm] was primary in the process of discovery. His method of excluding options, however, bore the greatest weight of his arguments of justification" (Churchill 1987 354).

23. The suggestion Weismann is here self-consciously rejecting with this denial is Strasburger's theory of "germinal return" (see Coleman 1965 150f).

24. In fact, Weismann would later (1896) develop a much more elaborate theory of the processes of competition, combination, and control between constituent elements of the germ-plasm that determined the course of growth and development (and the changing composition of the idioplasm) in each particular "track" or intraorganismic lineage of cells.

25. Although Weismann often writes that each cell is controlled by only a "single" determinant, other passages suggest that this is meant to imply only that such control is effected simply by a single *kind* of homologous determinant, with competitive and combinatorial propensities (e.g. differing degrees of "controlling force") among such homologous determinants and their constituent biophors acting to determine the ultimate characteristics of the cell (see GP chap. 1 sec. 4 and chap. IX sec. 3). Of course, Weismann also notes that organismic traits are often determined by the number, arrangement, proportions, repetition, rate of division, or other characteristics of the various cells constituting a particular structure, rather than by the ontogenetic fate of any particular cell.

26. More properly, the loss of substance in one direction simply ends a pre-existing "resistance to growth" in that direction by the organism's tissues, rather than being a "stimulus . . . in the ordinary sense of the word" (GP 129). Indeed, Weismann appeals to an analogous regenerative process to explain the unwelcome results of Driesch's sea urchin experiments (see Robinson 1979 181).

27. Even when Weismann considers the possibility that polymorphism in bees could be produced by differences in the amount or character of the nutrition with which they are supplied, he conceives of this as a matter of the determinants responsible for particular structures *only becoming active* when supplied with abundant nourishment. Weismann ultimately rejects this particular potential explanation in any case because each of the two forms he is considering has physiological structures that the other lacks (GP 376–377).

28. According to our own current theory, during meiosis the chromosomes of a single spermatocyte or oocyte are first doubled, producing twice the chromosome number of an ordinary somatic cell. In spermatogenesis, each spermatocyte then divides into four sperm, each with half the number of chromosomes in an ordinary cell. In oogenesis, however, only a single egg is formed from each oocyte, so this additional chromosomal material must be ejected during the transformation of an oocyte into the egg cell. These packets of surplus genetic material ejected during the maturation of the egg are referred to as the "polar bodies."

29. Perhaps most important in producing this recognition was Hertwig's careful point-by-point comparison of spermatogenesis and oogenesis, showing that meiosis involves a parallel sequence of unusual nuclear divisions that produces four sperm cells from a single spermatocyte, strongly suggesting in turn that each of the three polar bodies represented an undeveloped egg cell (see Churchill 1968 106–108, 1970 433).

30. This sense of 'expendable' is actually closest to the original military usage of the term, which designates supplies or equipment that are expected to be

used up, destroyed, or sacrificed in the course of a military engagement (e.g., ammunition) and therefore need not be listed on a certificate of expenditure.

31. This is certainly not to deny that the scientist's own intuitive distinction between more speculative and more empirically grounded theorizing might well be the beginning of wisdom about when the problem of unconceived alternatives does or does not pose a serious threat to eliminative practices of confirmation (cf. chap. 8). The point here, however, is that nothing so simple as dividing our theorizing into that which does and does not concern entities we can observe and/or detect will serve to delimit the scope of the problem. In chapter 7 we will also see (in connection with Stathis Psillos's attempt to defend realism from the historical record) why the scientist's intuitive distinction cannot serve all by itself as a reliable indicator of where the problem presents a significant challenge.

6

History Revisited

Pyrrhic Victories for Scientific Realism

Another such victory over the Romans, and we are undone.

—Pyrrhus, from Plutarch's *Lives*

6.1 Realist Responses to the Historical Record

The evidence we have seen from nineteenth-century theories of inheritance and generation strongly suggests that the problem of unconceived alternatives presents a serious prima facie challenge to scientific realism and to what we earlier called educated common sense about science. But it is not enough to show that there is a serious challenge in the offing, for realist philosophers of science have faced challenges grounded in the history of science before, and the case for the problem of unconceived alternatives will not be complete until we are sure that it is not rendered toothless by the responses they have offered. That is, although scientific realists have not yet seriously considered or confronted the problem of unconceived alternatives itself, recent years have witnessed a number of careful, thoughtful, and sophisticated realist responses to the original pessimistic induction over the history of science, and some of these responses, if successful, would give us reasons to doubt the significance of the problem of unconceived alternatives as well.

Realists have sometimes suggested, for example, that the claims of total failure for successful past theories on which the original pessimistic induction is founded are overblown, and that all genuinely successful past theories have turned out to be at least approximately true. If this were so, it would encourage us to discount not only the pessimistic induction itself but the problem of unconceived alternatives as well, for it would give us a reason to be confident in at least the approximate truth of our successful current theories, *even if we allow that there probably are indeed serious alternatives to them that remain presently unconceived.* In this way, a convincing analysis of the historical record intended to thwart the original pessimistic induction could give us grounds to embrace some form of scientific realism even while leaving unrebutted both the problem of unconceived alternatives and its implications for the reliability of eliminative inferences in particular epistemic contexts. Thus, showing that the problem of unconceived alternatives represents a serious threat to scientific realism will require us to consider what realists have had to say in response to existing challenges based on the historical record of scientific inquiry.

Accordingly, in this chapter and the next we will take up the most influential recent efforts by scientific realists to blunt or block the pessimistic induction by engaging the details of the history of science itself, including the first serious efforts to recruit those details to the realist cause. Despite the welcome sophistication and subtlety of much of this recent engagement, I hope to show not only that these arguments offer no convincing response to the problem of unconceived alternatives, but also that they do not even seriously compromise the original pessimistic induction to which they themselves are addressed! More specifically, I will argue that the most promising and influential realist replies to the historical challenge (including those of Clyde Hardin and Alexander Rosenberg, Philip Kitcher, Stathis Psillos, Jarrett Leplin, and John Worrall) ultimately manage to achieve only Pyrrhic victories for realism, that is, "defenses" of scientific realism that are forced to concede to the realist's opponent either just the substantive points that were in dispute between them or everything she needs for a convincing historical case against realism itself. Thus, I will try to show that *both* the problem of unconceived alternatives and the pessimistic induction itself survive even the best recent efforts to defend realism from the specter of the historical record.

6.2 Once More into the Breach: The Pessimistic Induction

Most recent efforts to engage the pessimistic induction begin from Larry Laudan's classic critique (1981, 1984b) of the explanationist defense of realism, so it is worth our time to revisit Laudan's influential discussion. Laudan's case focuses on the related questions of truth and reference for our successful scientific theories: he uses the historical record to cast doubt on whether successful contemporary scientific theories are (even probably or approximately) true and whether their central theoretical terms refer. While the matter of truth may seem straightforward enough, the question of reference may be unfamiliar to some readers. Although reference is a technical notion, the central idea is easy enough to grasp: in the study of language, a term is said to refer if it picks out something in the world and fail to refer if it does not. The term 'frogs' refers, then, because there is something in the world (namely frogs themselves) that is picked out by the term when we use it (e.g., to make claims about frogs). We judge that the term 'unicorns' fails to refer, because (we think) there is no actual object in the world that corresponds to it or is picked out by the term when we use it.[1] Closer to home, we tend to think that the term 'chromosome' is referential (and refers to chromosomes), while Weismann's term 'biophor' is nonreferential because there is not now and never was anything in the world that answers to this term: if our own contemporary biological theories are right, 'biophors' is like 'unicorns' and not like 'frogs.' Subsidiary discussion in the philosophy of language and linguistics centers on how and why particular terms come to refer to particular entities in the world (or to nothing at all), an issue that will become important in what follows.

Returning to Laudan, his wide-ranging discussion of the historical record challenges any number of realist commitments, from the claim that later theories typically preserve earlier ones as limiting cases or at least explain why they were successful when they were to the presumption that a theory whose central terms refer or one that is approximately true must be or is even likely to be successful, and the argumentative relationships here are intricate. If, for example, Laudan can show that neither reference nor approximate truth is even likely to ensure success (or, more modestly, that realists have failed to show that they are), then the reference and/or approximate truth of our best theories would hardly offer much of an explanation of their success at all, much less the best or only explanation promised by the explanationist or "miracle" argument for realism canvassed in chapter 1. Nonetheless, the lion's share of subsequent discussion has centered on Laudan's most simple and direct challenge to scientific realism. As a "confutation" of the explanationist argument's inference from the success of our current theories to their approximate truth and/or reference, Laudan points out any number of past scientific theories, all of which, he claims, were eminently successful in their day[2] but each of which has subsequently been judged either to be radically false and/or to have central theoretical terms that do not refer.

Laudan offers, for example, a famous (perhaps infamous) list of theories that he claims "involves in every case a theory that was once successful and well confirmed, but which contained central terms that (we now believe) were non-referring" (1981 122). To the cases of eighteenth and nineteenth century ether and subtle fluid theories he has already discussed (including the ether theory of nineteenth-century optics and electromagnetism, the caloric theory of heat, the theory of the electrical fluid, and theories of gravitational and physiological ethers), this list adds the following examples (adapted from Laudan 1981):

the crystalline spheres of ancient and medieval astronomy;
the humoral theory of medicine;

the effluvial theory of static electricity;
'catastrophist' geology, with its commitment to a universal (Noachian) deluge;
the phlogiston theory of chemistry;
the vibratory theory of heat;
the vital force theories of physiology;
the theory of circular inertia;
theories of spontaneous generation.

"I dare say," he continues, "that for every highly successful theory in the history of science that we now believe to be a genuinely referring theory, one could find half a dozen once successful theories that we now regard as substantially non-referring" (1981 123). And he ultimately adds to this list examples of successful past theories whose central terms *did* refer (by present lights) but are nonetheless now judged to be radically false, such as geological theories prior to the 1960s that denied lateral motion to the continents, chemical theories of the 1920s that assumed the atomic nucleus to be homogeneous, and late nineteenth-century physical/chemical theories assuming that matter was neither created nor destroyed.

Laudan tosses off this list of supposedly successful-but-false-and/or-nonreferential past theories pretty casually, along with the claim that it "could be extended *ad nauseum*" (1981 122), and defenders of realism have often begun their responses to the pessimistic induction by subjecting Laudan's list to well-deserved closer scrutiny. With at least some justification, these critics have suggested (cf. chap. 1) that a number of Laudan's examples can be safely ignored because they are drawn from sciences which were not yet "mature," had not passed a suitable "take off" point, or did not enjoy some distinctive *variety* of success that is exhibited by (some) contemporary theories and that really does demand the truth of a theory as its best or only explanation, even if empirical success in general does not. And if we reject such examples, they argue, the evidential base from which the pessimistic induction might be generated becomes severely restricted.

Of course, *bare* appeals to maturity and/or stricter standards of success threaten to undermine the explanationist defense of realism itself (cf. Laudan 1981 122–123). The point of that defense, after all, was that the empirical success of our scientific theories was supposed to demand the truth of those theories as its best or only explanation, and defenders of actual past theories often argued quite sincerely that the truth of those theories offered the best or only explanation of their own successes. If we now insist that further conditions must be satisfied in order to trigger this explanatory demand, we will need a principled rationale for why just that sort of success remains a reliable indicator of the truth of the theories that enjoy it, when others that equally excited our initial admiration and credence failed to do so. That is, we will want to know why the further condition should make such an important difference, besides the bare fact that it would immunize current theories from challenges based on the historical record.

Moreover, appeals to maturity risk becoming effectively tautological (cf. Laudan 1981 122) if the criterion required for "maturity" or a comparable epistemological privilege is simply that a theory be sufficiently in agreement with our *own* theoretical account of the natural world. Richard Boyd, for example, suggests that the "take-off point" for a "mature" science is "a point in the development of the relevant scientific discipline at which the accepted background theories are sufficiently approximately true and comprehensive" (1981 627).[3] For all these reasons, sophisticated realists recognize the need for a *precise, nontautological, and nonarbitrary* characterization of the kind of maturity or special variety of success that is supposed to distinguish current theories from their superseded predecessors, and we will consider specific attempts to formulate such a criterion below.

Furthermore, attempts to winnow down Laudan's list of successful-but-false-and/or-nonreferential past theories (or the historical record of which it is supposed to be just a sampling) seem to promise only limited relief from the historical challenge, for even the most severe recent critics of the pessimistic induction acknowledge that *some* of Laudan's examples of superseded theories were mature and/or successful in the special way(s) that are supposed to require the truth of the theory as their best or only explanation (see Kitcher 1993, 2001a; Psillos 1999; Leplin 1997; Worrall 1989, 1994).[4] For example, the phlogiston theory of chemistry defended in the eighteenth century by thinkers like Priestly and Cavendish posited the existence of a chemical "principle" ("phlogiston") that was supposedly expelled during the combustion of matter and which the surrounding air had a limited capacity to absorb. This theory offered systematic and unified explanations of a variety of chemical reactions (including calcination, reduction, and respiration, as well as the ordinary combustion of matter) and was able to explain many otherwise puzzling chemical and physiological phenomena, such as why combustion in a closed vessel would cease before the combustible material was consumed and why "dephlogisticated" air (air with phlogiston removed from it[5]) supported combustion and respiration better and longer than ordinary air. Likewise, the caloric theory of heat defended by Lavoisier and other eighteenth-century theorists offered systematic and comprehensive explanations of a wide range of thermodynamic phenomena (like phenomena of conduction and radiation and the role of heat in various chemical reactions) by postulating the existence of caloric, a "subtle" (or "imponderable") fluid that was transmitted from warmer bodies to cooler ones when they came into contact. As we will see in more detail in the next chapter, this theory was also able to make predictions of thermodynamic phenomena and effects not already known to occur. And the nineteenth-century wave theory of light proposed that light (and later, electromagnetism generally) itself consisted of a wave motion propagated in a substantival medium: the "ether," an extremely rarified but mechanical and material substance that generated elastic restoring forces on any of its constituent particles disturbed from equilibrium positions. This wave theory generated some of the most spectacular empirical successes of nineteenth-century science, including most famously Poisson's use of the theory to predict (incredibly) that there should be a bright spot of light at the center of the shadow of a perfectly circular disk. Indeed, this prediction was derived by Poisson as a *reductio ad absurdum* of Fresnel's formulation of the wave theory before its unexpected experimental verification by Arago turned this novel prediction into a dramatic source of confirmation for the wave theory itself! This came on top of the wave theory's notable successes in explaining and predicting other phenomena of reflection, refraction, interference, diffraction, and polarization. And as we've seen, we might fairly add one or more nineteenth-century theories of inheritance and generation to Laudan's list of past theories that at one time enjoyed the kinds of empirical support that have traditionally led their defenders to suppose that they must be true, but have nonetheless ultimately turned out to be false and/or nonreferential.

Of course Laudan's challenge cannot depend upon agreement with current theoretical science as a test of truth and/or successful reference, so his historical examples offer classically inductive evidence of the falsity of our own successful theories only if we have some further ground in each case for regarding these successful predecessors as definitively refuted, but the challenge need not be formulated in this way in any case. Laudan's ghostly historical procession might be taken to show that past theories have been quite successful in just the ways that impress contemporary realists while nonetheless making fundamental claims about nature that are simply *inconsistent* with those of present successful theories (as well as those of other past successful theories in the same domains). Since at most one theory in any such inconsistent set can be true, we are provided with abundant inductive evidence that empirical success cannot possibly be a reliable indicator of the truth about nature, *no matter what the truth is*: any general practice of inferring from such successes to the approximate truth and/or reference of the theories that enjoy them would have to routinely if not invariably lead us astray. Alternatively, we can see the realist inference from success to approximate truth and/or reference as *self-undermining*, for if the success of current theories leads us to conclude that they are approximately true and/or referential, this implies in turn that many past theories must have been radically false and/or nonreferential *despite* being successful, undermining our original ground for concluding that current theories are approximately true and/or referential in the first place.[6]

6.3 Reference without Descriptive Accuracy

One important strategy of recent realist reply to this historical challenge suggests that Laudan's classic defense of the pessimistic induction exaggerates the extent to which the central terms of the rejected past theories he considers should be judged nonreferential by present lights. Laudan seems to require, for example, that the descriptive claims theories make about their central posits be largely accurate in order for the central terms in those theories to successfully refer, and some critics have suggested that causal accounts of reference reveal this requirement to be prejudicially restrictive. On the causal accounts of reference these critics have in mind, a term refers to a particular object or set of objects out in the world not because the object(s) satisfy some description associated with the term, but instead in virtue of a causal relationship between the object(s) and the speaker who introduced the term into the language in the first place.[7] In the case of what are called "natural kind" terms, for instance, we can imagine a speaker in the distant past coming into direct causal contact with an actual sample of a substance, pointing it out to other members of her linguistic community, and announcing that she will call this substance by the name 'gold.' On a causal theory of reference, this act of "baptism" fixes the reference of the term by annexing it to the underlying constitution of the actual sample, so when the members of this linguistic community use the term 'gold' they manage to actually refer to gold (that is, all the stuff with atomic number 79) even if important aspects (perhaps even all) of the *descriptions* they associate with the term or their further *beliefs* about gold turn out to be profoundly mistaken (that it only occurs in the local mountains, that it is made up of elemental humors, that it is the tears of God, whatever). This explains how language users in the distant past could use terms like 'gold' (and 'water,' and 'tigers,' and so forth) to make false claims about actual gold and water and tigers, rather than making true claims about entities that existed only in their heads and/or being unable to make claims about objects in the actual world at all. Likewise, the term 'chromosomes' referred to chromosomes even as used by the nineteenth-century theorists who had radically false beliefs about the nature, constitution, and further properties of the chromosomes: the reference of this term was fixed, on this account, by a scientist's naming of the entities that he himself had seen through the microscope whether his further beliefs about those entities were correct or not.

Of course, terms naming the theoretical entities *posited* by a theory cannot be introduced into a language by pointing out *samples* of their intended referents, but the same general causal principle applies: on causal theories of reference, a theoretical term refers to whatever entities in the world actually cause the observable phenomena or events that led past theorists to introduce the term into their theories in the first place.

Thus, when theorists like Franklin and Ampère used the term 'electricity,' it referred to the *actual* electricity causing the phenomena (like sparks, lightning bolts, currents, and electromagnets) that led them to posit the existence of an unobservable physical magnitude responsible for these phenomena, despite their many false fundamental beliefs *about* (actual) electricity (i.e., that it was a fluid substance).

Causal theories of reference are not without their problems, but they certainly represent one of the live contenders for an account of how theoretical terms come to refer to particular objects in the world. How is this supposed to help the realist? The suggestion is that the viability of the causal account of reference makes it inappropriate for Laudan to assume that past theories must have been *descriptively accurate* in order for their central terms to be referential. Thus Hardin and Rosenberg argue that because "[o]ne permissible strategy of realists is to let reference follow causal role," realists are free to regard the central terms of even those theories they regard as radically mistaken to have been referential after all. For example, because we regard the electromagnetic field as playing the causal role attributed to the ether by earlier physical theories (such as the nineteenth-century wave theory of light), the realist may hold, they suggest, "that 'ether' referred to the electromagnetic field all along" (1982 613–614). And of course, if this constituted a convincing response to the pessimistic induction it would seem to count equally well against the problem of unconceived alternatives, for it would allow us to be confident that the central theoretical terms of our current successful theories will turn out to be referential even if serious and well-confirmed alternatives to those theories do in fact remain unconceived by us.

As a response to the historical record, however, the realist's appeal to causal theories of reference misses the forest for the trees. The sort of account envisioned by Hardin and Rosenberg secures a history of successful reference for terms in discarded theories only by explicitly divorcing their reference from the question of the accuracy of those theories and thus abandoning the specifically theoretical beliefs of the very sort for which the realist hopes to convince us to share her realism in the case of current theories. But this runs the realist afoul of what we might call the "trust" argument: after all, if the central terms in past theories are held to be referential *despite* the fact that the theories in which they are embedded repeatedly turn out to be radically misguided, then the historical record still suggests that we would be foolish to *trust* or *believe* either the theoretical accounts of inaccessible domains of nature offered by successful contemporary scientific theories or the descriptions associated with their central theoretical terms. And of course trusting the accounts of such domains and entities given by (some) current theories is just what the realist hoped to convince us to do! Thus, Hardin and Rosenberg achieve only a Pyrrhic victory for scientific realism: that is, a defense of realism that sacrifices the central tenets the realist sought to defend. Our confidence that the terms in our own successful theories are referential is purchased in a way that leaves us completely unable to be confident that their associated theoretical descriptions are also accurate, and that is just what was at issue all along!

Precisely because this shortcoming of Hardin and Rosenberg's account is so easily recognized,[8] it is surprising that a version of the same problem confronts more recently influential and sophisticated realist approaches to the reference of theoretical terms. Kitcher (1993 chap. 4), for instance, argues convincingly that the particular instances or *tokens* of a speaker's use of a natural kind term must be separated and her dominant linguistic intentions considered in assigning reference to them: thus the reference of some of Priestly's tokens of 'dephlogisticated air' are fixed by his intention to refer to air with the substance emitted in combustion removed from it, while the reference of other tokens is fixed by his intention to refer to the substance whose inhalation was rendering his breathing particularly light and easy or the substance he 'exploded together' with 'inflammable air' to produce either water or nitric acid. In the former cases, Priestly's tokens of 'dephlogisticated air' fail to refer (as there is no substance emitted in combustion in the way Priestly imagines), but the latter tokens instead refer to oxygen. Likewise, some of Fresnel's tokens of 'light wave' fail to refer because their reference is fixed by their theoretical description as oscillations of molecules of the ether, while the reference of others is fixed by Fresnel's dominant intention to talk about light, however it is in fact constituted, and therefore refer to electromagnetic waves of high frequency.[9] Kitcher concludes that claims of referential failure for past theories are overstated, and that the heterogeneity of the different modes of reference-fixing in different contexts of use permitted past theorists to make many referential and indeed true claims about the world.

But for all its success in rescuing the terms in discarded theories from blanket assertions of referential failure, Kitcher's approach runs afoul of the trust argument in just the same way that Hardin and Rosenberg's much simpler appeal to pure causal theories of reference did: Kitcher manages to rescue the reference of the central terms in discarded theories only *on those occasions* in which the user's dominant linguistic intentions explicitly eschew those specifically *theoretical* descriptions (like "the substance emitted in combustion" or "the oscillations of molecules of the ether") associated with her terms. But surely it offers little comfort to the realist if we insist that some tokens of terms like 'dephlogisticated air', and 'light wave' in rejected theories referred after all while admitting that the relevant theoretical accounts and descriptions of those entities were mistaken about virtually everything except the fact that the entities in question played some causal role in producing observable phenomena, for it is (once again) ultimately those theoretical accounts and descriptions which the realist hopes to defend in the case of current theories. Thus, Kitcher offers a welcome sophistication to our account of reference, but one that makes no progress whatsoever in defending realism from the historical challenge: his account shows how tokens of the central terms of past theories were referential (and past theorists were able to enunciate important truths) just where their being so (and doing so) did not depend upon those theories actually getting anything much right about the natural world. Likewise, the fact that the central theoretical terms of past theories referred *on those occasions in which their reference was fixed in this way* does nothing to undermine the challenge to the realist's belief in the descriptive accuracy of those theories posed by the problem of unconceived alternatives.[10]

By contrast, Stathis Psillos seems painfully aware that realists undermine their own cause in establishing mere referential continuity without descriptive accuracy for past theories. His own defense of realist reference (1999 chap. 12) is carefully constructed to avoid this particular pitfall by requiring at least some of the descriptive information associated with terms by our theories to be accurate in order for those terms to refer. It is especially revealing, then, that Psillos's account nonetheless concedes to his opponent all the resources she needs to make a convincing historical case against realism. Let us see why.

Psillos argues that a central problem for pure causal theories of reference is that they make referential success and continuity too easy to attain for theoretical terms. After all, there is always *something* in the world that really causes the observable phenomena that occasion the introduction of a theoretical term into the language, so on a pure causal theory it would seem that theoretical terms could never *fail* to refer: that is, a pure causal theory of reference would seem forced to deny our powerful linguistic intuition that terms like "caloric" or "gemmule" simply failed to refer to anything at all, and to insist implausibly instead that Lavoisier's talk of a caloric fluid draining into the spaces between the constituent molecules of bodies actually referred to molecular motion while Darwin's talk of "gemmules" cast off by our tissues and accumulating in the sex organs actually referred to genes. Likewise, a term like "phlogiston" would refer to oxygen, for oxygen is in fact the entity in the world responsible for the phenomena (of combustion, corrosion, respiration, etc.) that occasioned its introduction into our theoretical language.

Because pure causal theories make implausible referential continuity so easy to achieve in this way, Psillos argues that any convincing account of reference along these lines will have to be "causal-descriptivist"; that is, it will have to require not simply that a term refer to whatever causes the phenomena that occasion its introduction into the language, but in addition that some of the fundamental descriptions associated with the term actually be satisfied by the entity to which it refers. But not all the associated descriptions are equally important: he argues that "some descriptions associated with a term are less fundamental in view of the fact that the posited entity would play its intended causal role even if they were not true" (1999 297). The descriptions that really count, then, and which must be satisfied by an entity in order for a term to refer to it, are those making up what he calls the theory's "core causal description" of the entity in question: the descriptions that would *have* to be true in order for the entity to play the causal role the theory assigns to it. Thus, he argues, the term 'phlogiston' failed to refer because "there is nothing which fits a description which assigns to phlogiston the properties it requires in order to play its intended causal role in combustion" (1999 291), in particular "the property that it is released during the process of combustion" (1999 298).[11]

Of course, this account of the matter invites the realist to choose the core causal descriptions she associates with the central terms of past theories rather carefully, with one eye on current theories' claims about nature, so there is more than a whiff of ad hoc-ery about the proposal. But even if we set this worry aside and allow the realist to delicately titrate the core causal descriptions she associates with the crucial terms in successful past theories so as to render them referential by the lights of current theories, Psillos's victory will *nonetheless* remain a Pyrrhic one. To see why, let us look more closely at the one case of reference Psillos examines in detail: the luminiferous ether of nineteenth-century optics and electromagnetism.

It is important to Psillos's account that the term 'ether' in nineteenth-century wave theories of light and electromagnetism turn out to be referential, for (in contrast to Kitcher) he acknowledges that the postulation of the ether (as a "dynamical structure" serving as a carrier for light waves) played a crucial role in the successes of those theories. He argues that this demand creates no problem for the realist, however, because our own term 'electromagnetic field' shares the core causal description associated with the ether of nineteenth-century optics, and so, he claims, the term 'ether' referred to the electromagnetic field itself.[12] Although current physical theories regard the beliefs of nineteenth-century optical and electromagnetic theorists about the *nature and constitution* of the ether to have been quite radically and fundamentally misguided, Psillos insists that such beliefs did not form any part of the actual core causal description of the term 'ether'. That is, the false beliefs of nineteenth-century theorists about the mechanical, material, and substantival character of the ether were not part of the description that an entity would have to satisfy in order to play the causal role assigned to the ether by nineteenth-century wave theories of light and electromagnetism. And of course, it would seem that we have much less to fear from the problem of unconceived alternatives if the historical record shows that we can safely rely on the accuracy of at least the core causal descriptions associated with the central theoretical terms in our successful theories.

The reason Psillos's victory remains a Pyrrhic one for the scientific realist is that this case for the referential status of central terms in successful past theories simply invites from the historical record a renewed form of the pessimistic induction itself, this time concerning our ability to distinguish (at the time a theory is a going concern) which of our beliefs about an entity are *actually* part of its core causal description. The core of the problem is that nineteenth-century theorists themselves strenuously disagreed with *the very assessment Psillos offers* of what would have to be true of an entity in order for it to play the ether's causal role, that is, with his very claim about which descriptions of the ether enter into its core causal description. More specifically, they considered whether the ether could play the causal role ascribed to it in propagating light and other electromagnetic waves without consisting of a material medium of some kind *and explicitly denied that it could*! Maxwell himself concludes *A Treatise on Electricity and Magnetism* (1955 [1873]) with the following resolute insistence that the ether must be a material or substantival medium in order to play the causal role of propagating energy waves:

> If something is transmitted from one particle to another at a distance, what is the condition after it has left the one particle and before it has reached the other? If this something is the potential energy of two particles, as in Neumann's theory, how are we to conceive this energy as existing in a point of space, coinciding neither with the one particle nor with the other? *In fact, whenever energy is transmitted from one body to another in time, there must be a medium or substance in which the energy exists after it leaves one body and before it reaches the other*, for energy, as Torricelli remarked, 'is a quintessence of so subtile a nature that *it cannot be contained in any vessel except the inmost substance of material things*'. Hence all these theories lead to the conception of a medium in which propagation takes place, and if we admit this medium as an hypothesis, I think it ought to occupy a prominent place in our investigations, and that we ought to endeavor to construct a mental representation of its action, and this has been my constant aim in this treatise. (1955 [1873] Vol. II 493, my emphasis)

Maxwell's positive conception of the field is a matter of some legitimate controversy, but here he quite explicitly denies that anything *besides* a material substance is capable of playing the causal role he assigns to the ether: he insists that to avoid the patent (or at least inconceivable) absurdity of "energy . . . existing in a point of space" we must recognize the existence of a material and mechanical medium filling the space between bodies, for energy "cannot be contained in any vessel except the inmost substance of material things." Thus, Maxwell self-consciously insisted that the core causal description of the ether *included* the very beliefs about the ether's material and mechanical character that Psillos grants are not satisfied by the modern conception of the electromagnetic field as a distinct entity ontologically on par with matter itself.[13]

Thus, even if Psillos can convincingly argue that the *actual* "core causal description" of the ether does not include claims about its material or mechanical character, he will be forced to concede that the carefully considered judgments of leading scientific defenders of the theory concerning which of the descriptions associated with its central terms must be satisfied by an entity in order for it to play the causal role associated with the term (i.e. which features figure in the actual core causal description) have proved to be unreliable. What this suggests, of course, is that we cannot rely on our *own* judgments about which of the descriptions we associate with our own terms are *genuinely* part of their own core causal descriptions.[14] And if we cannot retain any confidence that the core causal descriptions *we* associate with the central terms of our own theories are indeed correct, then the challenges to realism posed by both the pessimistic induction and the problem of unconceived alternatives reassert their full force. Here the realist's victory over the pessimistic induction takes on a Pyrrhic character not because we can accept it while still denying that we can trust anything our current theories say about nature, but instead because it leaves us with what we might call a *failure of discrimination*: it allows us to trust only *some* of what current theories tell us about the natural world (e.g. the *real* "core causal descriptions" of the theoretical terms in our successful theories, whatever they turn out to be) while leaving us completely unable to be confident in our ability to discern just which parts of our theories *actually* constitute this privileged class of theoretical claims.

Once this underlying problem with Psillos's strategy is recognized, further supporting historical examples are relatively easy to identify. Indeed, we have already seen one further example in Weismann's insistence that the germinal or hereditary materials must be separated and reduced in each cell-division until each cell contains only the tiny amount necessary for determining its own characteristics. Recall that Weismann's argument for this germinal specificity explicitly claims that there is simply no other way to account for the fact that different parts of the body possess different characteristics, and, in particular, insists that the cells of the body could not possibly be heterogeneous if each cell carried precisely the same hereditary materials:

> As the thousands of cells which constitute an organism possess very different properties, *the chromatin* which controls them *cannot be uniform; it must be different in each kind of cell.*
>
> The chromatin, moreover, cannot *become* different in the cells of the fully formed organism; the differences in the chromatin controlling the cells must begin with the development of the egg-cell, and must increase as development proceeds; for otherwise the different products of the division of the ovum could not give rise to entirely different hereditary tendencies. This is, however, the case. Even the first two daughter-cells which result from the division of the egg-cell give rise in

> many animals to totally different parts. . . . The conclusion is inevitable that the chromatin determining these hereditary tendencies is different in the daughter-cells. (1893 [1892] 32, all emphases in original)

The point here is not simply that Weismann made (by current lights) some mistakes about the constitution of the germ-plasm, still less is it important whether these false beliefs prevented particular tokens of terms like 'chromatin' or 'germ-plasm' from referring on particular occasions of use. Rather, the point is that Weismann argued quite explicitly that *no entity could possibly play the causal role that he assigned to the hereditary material without being separated and parceled out differently to different cells*, offering further evidence of our historical unreliability in judging which theoretical descriptions are genuinely part of what Psillos calls a term's core causal description.

Psillos's strategy is similarly undermined by cases in which the term in question is presently judged to be uncontroversially *non*referential. Consider Antoine Lavoisier's claim, defending the caloric fluid theory of heat in his 1785 "Memoir on Phlogiston," that:

> One can hardly think about these [thermal] phenomena without admitting the existence of a special fluid [whose accumulation causes heat and whose absence causes cold]. It is no doubt this fluid which gets between the particles of bodies, separates them, and occupies the spaces between them. Like a great many physicists I call this fluid, whatever it is, *the igneous fluid, the matter of heat and fire.* (Lavoisier 1785 as translated in Donovan 1993 171, original emphasis, translation modified)

This confidence that *only* a subtle fluid could play the role in causing thermal phenomena that he assigns to caloric appears even more explicitly in the course of Lavoisier's later work:

> It is difficult to conceive of these phenomena without admitting that they are the result of a real, material substance, of a very subtile fluid, that insinuates itself throughout the molecules of all bodies and pushes them apart. . . . This substance, whatever it is, is the cause of heat, or in other words, the sensation that we call heat is the effect of the accumulation of this substance. . . . (*Traité de Chimie*, in Lavoisier (1965 [1743–1794], vol. I 1–3, my translation)

Thus Lavoisier's nonreferential account of caloric shows, no less than Maxwell's supposedly referential discussion of ether and Weismann's uncontroversially referential account of the germ-plasm or chromatin, that we have historically been unreliable (again, by present lights) in judging which descriptions must be satisfied by an entity or what characteristics it must have in order for it to play the causal role assigned to it by a particular theory. The significance of this point is, of course, entirely independent of the fact that 'caloric' does not refer: what matters is not that Lavoisier was wrong about the need to posit caloric, but rather that he was mistaken in his insistence that nothing *besides* a subtle fluid could play the role in causing thermal phenomena that he assigns to caloric (but that contemporary theorists assign instead to molecular motion). And as this point illustrates, it will not help to respond that Psillos is wrong about the need to defend the claim that the term 'ether' was referential: if history reveals us to be repeatedly mistaken in judging which of our beliefs about posited entities are really part of their "core causal descriptions" (whether the names of those entities are ultimately judged to be referential or not), this will prevent us from knowing which are the beliefs on which we can safely rely concerning *whatever* entities are posited by our current successful theories. Thus, the problem described here for Psillos's "core causal description" strategy in no way depends on the truth of his (contentious) claims that the postulation of the ether played a crucial role in the success of nineteenth-century theories of light and electromagnetism and/or that the term 'ether' actually referred to the electromagnetic field.

Like Kitcher's then, Psillos's defense of the referential status of the central terms in past successful theories leaves us unable to be confident that any *particular* theoretical descriptions we associate with a referring term (even ones we presently regard as absolutely central and/or indispensable to fulfilling its causal role) will be retained in the further development of theoretical science, and thus comes only at a price realists cannot afford to pay. And for this same reason, Psillos's Pyrrhic victory does nothing whatsoever to mitigate the challenge posed by the problem of unconceived alternatives.

6.4 Diluting Approximate Truth

Related difficulties afflict the complementary realist strategy of suggesting that Laudan is too quick to deny that his examples of superseded historical theories were (at least approximately) true. Once again, if this constituted a convincing avenue of reply to the pessimistic induction, it would call the significance of the problem of unconceived alternatives into question as well: if all or nearly all genuinely successful past theories have turned out to be approximately true, we would have good reason to think that contemporary successful theories are (probably, approximately) true, *notwithstanding* the probable existence of serious theoretical alternatives to them that remain unconceived by contemporary scientists (just as in the case of past theories).

There is a further complication here, however, in that Laudan takes advantage of an argumentative shortcut in making his case against the approximate truth of rejected theories: he assumes explicitly that the failure of a theory's central terms to refer (in a sense that requires descriptive accuracy) ensures that the theory is not approximately true, arguing that "the *realist would never want to say that a theory was approximately true if its central theoretical terms failed to refer*. If there were nothing like genes, then a genetic theory, no matter how well confirmed it was, would not be approximately true" (1981 121, original emphasis).[15] Hardin and Rosenberg (1982) protest, insisting that a case like classical Mendelian genetics illustrates how a past theory might have central terms that fail to refer even by Laudan's descriptive criteria (for they insist, perhaps implausibly, that contemporary genetics recognizes nothing like a Mendelian gene) but nonetheless earn a judgment of approximate truth by today's practitioners. Thus, they claim, Laudan's case relies inappropriately on the reference of central terms as a minimal condition for approximate truth.[16]

Laudan's own response to this objection asks by what right Hardin and Rosenberg "take contemporary theories as benchmarks of what there is and how it behaves" once they have granted that a successful theory (like Mendel's) may be very wide of the ontological mark and thereby undermined the realist's abductive argument itself (1984a 159). But this reply risks missing the point of the objection: Hardin and Rosenberg can respond that they are simply showing how a current theory, *if* true, could ground the judgment that a particular past theory was both nonreferential and approximately true, thereby invalidating Laudan's argumentative shortcut from failures of reference to failures of approximate truth, and thus undermining the case he makes against the explanationist defense of realism in the first place.

More effective, therefore, is to ask what weight this sort of objection to Laudan's argumentative shortcut is ultimately supposed to carry. Even if we grant Hardin and Rosenberg that the relationship between Mendelian and contemporary genetics illustrates how one theory can be (by another's lights) both nonreferential and approximately true, *this is surely not what the verdict of our own actual contemporary theories would be concerning most, if not all, of the other theories on Laudan's historical hit parade*. If, as the realist would have it, our contemporary theories are true, then we are inclined to insist that the many eminently successful theories included on Laudan's list (including the phlogiston theory of chemistry, the caloric theory of heat, and the theory of light and/or electromagnetism as a wave motion propagated through a material or substantival ether) are *not* in fact true, not even approximately so: the relationship between current theories and those on Laudan's list is simply not, in general, the one that Hardin and Rosenberg claim obtains between Mendelian and contemporary molecular genetics. Furthermore, the claim that these successful past theories are not even approximately true by contemporary lights is indeed strongly *supported* by the fact that current theories hold there to be nothing like the central posits of those superseded theories, even if this fact is not alone sufficient to guarantee their radical falsity by contemporary lights.

Of course Hardin and Rosenberg might instead have in mind some sense of "approximate truth" in which the literal truth of current theories is indeed consistent with the approximate truth of such classic success stories as the phlogiston theory of chemistry, the caloric theory of heat and wave theories of the optical and electromagnetic ether, but if so they will quickly find themselves back in the jaws of the trust argument. This is because the "approximate truth" of past successful theories in *any* sense that is consistent with being as fundamentally and profoundly mistaken about the constitution of nature as these famous predecessors were by the lights of current theories simply does not cut against the suggestion that the historical record shows why it would be a mistake to trust or believe the theoretical accounts of nature they offer or that doing so would have routinely led us badly astray in the past. Thus the realist will win a battle over something she is pleased to call "approximate truth," but again lose the war over realism: she will again be a victim of the trust argument, for her opponent's skepticism about the accounts of nature offered by current successful theories will rightfully survive her concession that those theories may well be "approximately true" in the attenuated sense the realist has managed to defend for successful past theories.

This same point undermines Kitcher's suggestive analogy between the response of modest scientific realists to the historical record and "that of the author who confesses in her preface that she is individually confident about each main thesis contained in her book but equally sure that there's a mistake somewhere" (2001a 171; see also 2001b 18–19). The analogy is clever, for it invites us to see opponents of realism as simply carping over our inability to attain an unreasonable standard of accuracy in what is admittedly a difficult business. Nonetheless, even when we restrict our attention (as Kitcher insists) to those historical cases of past theories grounding "predictions and interventions that were numerous, diverse, and hard to achieve" (2001b 19), the analogy proves to be seriously misleading, for contemporary theories are surely not rightly thought to hold such successful predecessors as Newtonian mechanics, nineteenth-century wave theories of optics and electromagnetism, or the caloric theory of heat to have been mistaken simply in matters of minor detail comparable to a misplaced footnote, a speculative musing, or even an overstated conclusion.[17] Instead, they hold these illustrious predecessors to have been deeply and thoroughly mistaken in their most central claims about the constitution and/or operation of nature, a situation more analogous to the author having been fundamentally mistaken in the principal thesis of her book or the central contentions she was concerned to advance. Thus, diluting approximate truth down to the point where all genuinely successful past theories can qualify gives us no reason to think that the admitted "approximate truth" of *contemporary* theories in this attenuated sense would in any way mitigate the challenge to believing them that is posed by the likely existence of unconceived theoretical alternatives. We may once again simply grant the proffered defense of realism, but still answer the crucial question—"Should we trust or believe what our own best scientific theories tell us about what things are like in otherwise inaccessible domains of the natural world?"—with a resounding "no."

Other recently influential efforts to defend realism from the historical record turn out to dilute approximate truth in ways that are similarly self-defeating. Jarrett Leplin's extended defense of realism (1997), for example champions the inference from a theory's success in making *novel* predictions (in a precisely specified sense of novelty[18]) to what he calls its "partial truth" (1997 127). Leplin modestly aims to defend only what he calls Minimal Epistemic Realism: the claim that there *are* epistemic conditions that would warrant a realist attitude toward a theory, not that any present theory actually satisfies such conditions. But the partial truth Leplin argues we can infer from a theory's novel predictive success is nonetheless too meager to render the prospect of this inferential entitlement anything more than another Pyrrhic victory for scientific realism.

This is because Leplin rightly sees that he cannot infer any particular degree, kind, or respect of partial truth (in the sense of representational accuracy[19]) from any particular degree, kind, or respect of success in novel prediction. He grants that he is "vague by default as to how much novel success merits what level of confidence in representational success," explicitly denies that "the amount of novel success provides a measure of the degree of representational accuracy achieved," and acknowledges that novel predictive success does not warrant attributing "a particularly high level of accuracy, because we have no way to determine what forms of inaccuracy might be irrelevant to the observable situation" (1997 128). But with no inferential connection between degrees or kinds of novel predictive success and degrees or kinds of representational accuracy, we will never be able to say anything more about any theory (even a merely possible future theory) than that it enjoys "some" degree of representational accuracy, no matter how much novel predictive success it has enjoyed. And this (unimprovable) claim is trivially satisfied: after all, Aristotelian mechanics, Creationist biology, caloric thermodynamics, and phlogistic chemistry all enjoy "some" degree of representational accuracy, too—none of these theories was wrong or misleading (by present lights) about absolutely everything, not even everything fundamental.[20]

Thus, although he intends to eschew the traditional realist commitments he regards as indefensible, Leplin cannot, I suggest, defend even a minimal realism worth the name with such a feeble connection between novel predictive success and representational accuracy or partial truth. Instead his account dilutes the notion of "partial truth" to the point that the historical record of such partially true theories supports rather than opposes the antirealist's fundamental skepticism about the descriptive or representational accuracy of current (and even future) successful theories.[21] Once again, then, the concession that our own theories with successful novel predictions to their credit may or even must be "partially true" in Leplin's sense will not draw the sting either from the traditional pessimistic induction or from the problem of unconceived alternatives itself.

Perhaps for this reason, Leplin sometimes seems tempted to reverse himself and assert a more fine-grained connection between novel predictive success and a particular degree or kind of representational accuracy: responding to the pessimistic induction, he suggests that the fundamental theoretical mechanisms employed by theories that enjoyed novel predictive success have not been overturned by subsequent developments (1997 145). But he is fully aware of the problem that such a suggestion creates: the classic textbook example of novel predictive success—the prediction of the Poisson bright spot—was made by a theory now regarded as radically misguided, namely, Fresnel's formulation of the wave theory of light, with its conception of light waves as oscillations of the molecules of a material ether. Therefore, Leplin's "direct" response to the pessimistic induction (1997 146f) goes on to appeal hopefully to a further suggestion made by Kitcher that we might call the strategy of *selective confirmation*: we can show, Kitcher suggests, that those parts or aspects of rejected theories actually responsible for their successes have been preserved in or ratified by subsequent theorizing about nature, and these were the only components of past theories, he argues, for which we had any real confirmation in the first place.[22] That is, Kitcher argues (as does Psillos 1999) that past practitioners had genuine confirmation only for those *parts* of past theories that have turned out to be *true*.[23]

This strategy of selective confirmation seems to represent the most promising hope for any effort to disarm the problem of unconceived alternatives (or the pessimistic induction itself, for that matter) by suggesting that our past successful theories have turned out to be approximately true after all. Any number of past theories have enjoyed truly impressive empirical successes, and efforts to dilute the notion of "approximate truth" (or "partial truth") to the point that it will include all such successful theories run afoul of the trust argument discussed above: they allow us to simply accept the proffered defense while rightfully persisting in our skepticism about the fundamental claims made about the natural world by our own best current theories. Accordingly, it seems that any successful defense of realism along these lines will be forced to separate past theories into constituent elements and insist that those parts genuinely confirmed by the successes of those theories have been preserved in and verified by our further theorizing. We now turn, therefore, to a more detailed assessment of the strategy of selective confirmation itself.

Notes

1. I here pass over the fact that "unicorn" can sometimes refer to an imaginary, notional, or intentional entity in hopes that the relevant contrast between

referring and nonreferring terms is already tolerably clear to anyone who is sufficiently clever and/or familiar with the issues to point this out.

2. With some reservations I will here follow the regrettably common practice of characterizing the relevant sorts of success at a painfully vague or intuitive level, though I am happy to accept the appealingly commonsensical account of such success (grounded in the potential for solving practical problems of prediction and intervention) offered by Kitcher (2001a 166–167). As a general matter, however, defenders of the pessimistic induction have resisted tying themselves too closely to *any* particular conception of scientific success, seeking instead to argue that there is (at most) a difference of degree and not in kind between present and past theories with respect to *whatever* sorts of success realists suppose could only be explained by the truth of the theories that enjoy them (see e.g. Laudan 1981).

3. By contrast, Clyde Hardin and Alexander Rosenberg propose instead that theories past the take-off point are "both comprehensive and robust, i.e., supported by convergent lines of independent argument" (1982 610). While this proposal is not tautologically empty in the way that Boyd's is, the historical examples we have examined in earlier chapters already illustrate why it will also not do much to help the realist, as radically false past theories have been repeatedly able to satisfy this criterion of maturity. As we noted in chapter 3, for example, it was perfectly reasonable for nineteenth-century teleo-mechanists to regard the existence of developmental or formative vital forces as supported by multiple lines of independent evidence converging from diverse sources.

4. Thus, the debate over maturity and/or special varieties of success concerns not whether the historical record supports some version of the pessimistic induction but instead the *strength* of the support it offers. For the interesting suggestion that both the pessimistic induction and the explanationist defense of realism depend on the fallacy of ignoring base rates which are themselves impossible to estimate, see Magnus and Callender (2004).

5. Dephlogisticated air was generated by heating the "calx" of a metal in a closed vessel, producing the metal itself by—according to the theory—absorbing phlogiston from the surrounding air.

6. This last formulation of the challenge is essentially that offered by Worrall (1994 334).

7. On such theories, successful reference in a linguistic community also depends on causal processes of transmission of the term from one user to another, but we may safely ignore this complication for present purposes.

8. See Worrall (1989, esp. 116–117), Laudan (1984a, esp. 161), and Psillos (1999, esp. chap. 12).

9. This strategy of analysis is further developed in connection with a general causal-descriptivist account of reference for natural kind terms in Stanford and Kitcher (2000).

10. Kitcher's further argument (1993 chap. 5) that those *parts* of superceded theories enabling them to be successful were also true (see also Psillos 1999) would seem to require that at least some of the central terms of past theories referred successfully even on occasions when their reference was fixed in what he calls the "descriptive mode" and the descriptions in question were theoretical (and conversely, that where particular uses of past terms were

sufficiently infected with false theory as to be nonreferential, they were not involved in or responsible for those theories' successes). In the next chapter I will argue that this argumentative strategy also secures only a Pyrrhic victory for the realist, although for a very different reason.

11. In contrast to the position defended in Stanford and Kitcher (2000), I now find it more natural to think of such considerations not as establishing it as a constant and unalterable fact that the term 'phlogiston' failed to refer, but instead as grounding our later *decision* to treat 'phlogiston' as having no referent on particular occasions in *interpreting* the language of an earlier community, but this difference does not matter for present purposes.

12. Actually, Psillos's claim seems better suited to the early twentieth-century conception of the electromagnetic field than the contemporary one: it is at least contentious to describe the electromagnetic field recognized by contemporary quantum electrodynamics as one in which light waves *propagate* at all (cf. Psillos 1999 296). But I will not rely on this point in what follows: instead I will show that Psillos's realist victory is hollow even if we treat this early twentieth-century conception of the field as our own.

13. Of course, the contemporary conception of the electromagnetic field ascribes to it a well-defined (local) mass-energy content and it may therefore be said to qualify as a "material substance" in some current sense of this term. (My thanks to David Malament for helpful discussion on this point.) But the sense in question is far removed from the requirements Maxwell had in mind in insisting that the ether must be a "medium or substance" or denying that energy can reside anywhere "except the inmost substance of material things" (and Psillos seems to grant as much in recognizing that nineteenth-century theorists' beliefs about the nature and constitution of the ether have turned out to be radically mistaken). Thus, we must understand Maxwell as insisting that nothing besides a material or mechanical substance as *he* understood these notions could play the causal role he assigned to the ether, and it is the inaccuracy of this judgment (by modern lights) that suggests a renewed pessimistic induction concerning our ability to discern just which parts of the descriptions we associate with terms in our successful theories are really part of their "core causal descriptions." Perhaps it is also worth noting that Maxwell's confidence in the need for a material medium seems to depend on his failure to conceive of anything like the electromagnetic field of later physical theory.

14. Although Psillos admits that there is an element of rational reconstruction in settling on the appropriate core causal descriptions for terms in past theories, this simply glides over the real problem we have noted: without justifiable confidence in our ability to accurately specify core causal descriptions for theoretical entities, we don't know which features of our *own* theories are rightly included in the core causal descriptions associated with *their* central terms, and thus have no way to pick out which of our own actual beliefs we can trust.

15. Laudan avails himself of this shortcut in part because he argues that realists have failed to provide any account of approximate truth on which the presumption that the approximate truth of a theory implies or entails its likely success can be defended. He suggests that typically this presumption is uncritically assumed to follow from the unobjectionable fact that a perfectly true (and, we might add, complete) theory would be perfectly successful—an inference that

he insists is patently invalid—and he challenges realists to provide an analysis of approximate truth on which the presumption of success for approximately true theories is defensible. Thus, Laudan cannot afford to pin his argument on any particular conception of approximate truth itself.

16. Of course, as noted above, not all of Laudan's examples proceed in this fashion: he takes explicit notice of several theories whose central terms *did* refer, but which he suggests were nonetheless not approximately true (1981 123–124).

17. Cf. Worrall (1994 340): "A displacement current in a *sui generis* electromagnetic field and a mechanical vibration transmitted from particle to particle are more like 'chalk and cheese' than are real chalk and cheese." And a little less colorfully, "[o]ne would be hard pressed to cite two things more different than a displacement current, which is what [the] electromagnetic view makes light, and an elastic vibration through a medium, which is what Fresnel's theory had made it" (Worrall 1989 116). If this is an error of detail, then the devil is indeed in the details.

18. Novel predictions are traditionally regarded in the philosophy of science as the predictions of previously unknown, unexpected, or otherwise unlikely phenomena (see Barrett and Stanford 2005), so it is worth noting that Leplin's conception of novel predictive success is highly idiosyncratic: for Leplin, a result counts as a novel prediction for a theory just in case there exists some adequate rational reconstruction of the reasoning leading to that theory which does not appeal to even a qualitatively generalized description of the result and no other extant theory can predict even a qualitatively generalized description of the result (for details, see his 1997 chap. 3). As far as I can tell, the first of these conditions is trivially satisfied for every implication of any theory that enjoys at least two sources of support (for then the theory could have been developed without any one of its supporting lines of evidence), and the second concerns the existence of competitors to the truth of a theory as explanations for the empirical phenomena rather than the novelty of a given result for a given theory at all. Leplin explicitly intends, however, to eschew what he calls our "intuitions" about novelty in favor of trying to pick out just those forms of empirical success for which he thinks the truth of a theory that enjoys them provides the only possible explanation. For an argument that truth is nonetheless neither the only nor the best explanation of even Leplinian novel predictive success, see Stanford (2000). Worrall's (1989) conception of novel predictive success is far more intuitive, but (as he notes) cannot be used to defend traditional realist commitments because it is regularly exhibited by theories that are subsequently rejected; Worrall's own 'structural' realist conclusion will be challenged in chapter 7.

19. Leplin articulates two quite distinct senses of partial truth. The 'pragmatic' sense seeks to capture what practicing scientists commit themselves to in accepting theories, both prospectively (contrasting the partial truth of present theories with falsity) and retrospectively (contrasting the partial truth of past theories with unqualified truth), while the 'metaphysical' sense of partial truth is understood in terms of representational accuracy and is what we are supposedly entitled to infer from a theory's success in novel prediction. It is only this second, metaphysical sense of partial truth that will concern us here. My thanks to Jarrett Leplin for clarifying this and other aspects of his work in correspondence, though he would not be at all satisfied by the conclusions I reach.

20. For example, while none of Creationist biology's causal or explanatory mechanisms are accepted in current scientific theory or practice, its theoretically motivated division of organisms into species is nearly identical to the leading general approach in contemporary evolutionary theory (that is, the Biological Species Concept). The other three examples arguably enjoy some degree of representational accuracy by present lights even at the level of causal and explanatory mechanisms: for example, one of the important respects in which Kuhn famously suggests that "Einstein's general theory of relativity is closer to Aristotle's [mechanics] than either of them is to Newton's" (1996 [1962] 206–207) is presumably that general relativity (like Aristotelian mechanics) recognizes gravitational motion as itself a 'natural' state of motion (that is, along a 'straight' trajectory in curved spacetime) not requiring further causal explanation, rather than a deflection from natural (inertial) motion as in Newtonian mechanics.

21. This also seems the appropriate response to Larry Sklar's contention (2000 esp. sec. 4.1) that our best current theories should be viewed in light of history as "on the road to truth," "pointing toward the truth" or "heading in the right direction." It is far from obvious, however, that Sklar would disagree with this response and his central contention is that the most interesting and important issues are simply obscured at this level of abstraction in any case.

22. Leplin is particularly concerned, of course, with the suggestion that we have had selective confirmation only for those parts of past theories that were responsible for their successes *in novel prediction* (of his idiosyncratic variety).

23. In private correspondence, Leplin has indicated that he is disinclined "to try to turn what is clearly a problem for realism into a positive argument" in the way I suggest here, because even when the theoretical mechanisms responsible for novel predictive success survive in subsequent theories they are sometimes "so radically reconceived . . . that I do not hold much hope for founding upon their retention greater specificity as to [what] descriptive content novel success warrants a commitment to." He thus rejects the response I explore here and will have to accept instead that success in novel prediction warrants only an inference to "some" degree of representational accuracy, an inference which I have argued above trivializes the realism it seeks to defend.

7

Selective Confirmation and the Historical Record

"Another Such Victory over the Romans"?

Vanity plays lurid tricks with our memory.

—Joseph Conrad, *Lord Jim*

7.1 Realism, Selective Confirmation, and Retrospective Judgments of Idleness

As we saw in the preceding chapter, the best remaining hope for defending realism from the challenges of history seems to lie with the strategy of "selective confirmation." On this strategy, we defend only some *parts* or *components* of past theories as responsible for (and therefore confirmed by) their successes, while abandoning others as idle, merely presuppositional, or otherwise not involved in the empirical successes those theories managed to achieve, and therefore never genuinely confirmed by those successes in the first place. Hopeful appeals to this strategy can now be found in the writings of any number of scientific realists[1], but the strategy itself has been developed with greatest sophistication and attention to the details of the historical record by Philip Kitcher (1993) and Stathis Psillos (1999).

Kitcher and Psillos both acknowledge that the history of science includes genuine examples of mature theories that enjoyed impressive empirical successes, but each also insists that the pessimistic induction paints its history of "successful" but "false" theories with too broad a brush: it ignores, they argue, the fact that only some parts of these superseded theories were involved in the successes they enjoyed. As Kitcher puts it, "[n]o sensible realist should ever want to assert that the *idle* parts of an individual practice, past or present, are justified by the success of the whole" (1993 142), and he suggests as an illustrative case that the successes of those geological theories that labored under the constraint that lateral movements of the continents not be permitted came in areas in which this constraint was simply *irrelevant*. And Kitcher and Psillos each go on to suggest that as a perfectly general matter the features of superseded theories on which their successes genuinely depended are the very ones that have turned out to be retained in later theories and ratified by further inquiry. If this claim can be convincingly defended, it would seem that we have much less to fear from the problem of unconceived alternatives (not to mention the pessimistic induction itself), for we might then rest confident that at least those parts of our *own* theories genuinely responsible for and genuinely confirmed by the successes of those very theories will be retained in any further accounts of nature that we develop, even if once again we allow that there probably are indeed serious and well-confirmed alternatives to present theories that remain unconceived by us. Indeed, if we can show that those parts of past theories genuinely responsible for their successes have routinely or invariably been preserved in the course of subsequent theorizing about nature, then the historical record becomes a powerful testament to the *reliability* of a sufficiently *selective* inference from the success of (some parts of) our theories to their truth.

This selective confirmation approach also proves central to Kitcher's more recent defense of realism (2001a, 2001b) using what he calls the "Galilean Strategy" of generalizing the inference from success-to-truth in everyday contexts like card games and attempts to use the subway system (where its reliability can be checked) to that of theorizing about the natural world. Kitcher's defense of this generalization depends, as he recognizes, on establishing convincingly that theories' "past successes stem from parts of the theories that are approximately correct" (2001a 170), that is, on the ability of the strategy of selective confirmation to turn apparently failed instances of the success-to-truth inference strategy from the historical record of scientific inquiry into successful ones. Otherwise we have compelling reasons to doubt that the reliability of the everyday success-to-truth inference survives its Galilean importation into the quite different context of scientific theorizing, and Kitcher's argument turns on the suggestion that there is no reason to expect such a difference.

Laudan anticipates, at least in a general way, the temptation for realists to restrict confirmation only to parts of past theories, and argues in reply that scientific realists cannot afford to be anything less than complete holists in matters of confirmation. That is, he suggests that without a theory of evidential support on which "evidence for a theory is evidence for *everything* the theory asserts" (1981 116, original emphasis), realists cannot allow a theory's successes to confirm its deeply theoretical claims and posits along with its observational implications:

> After all, if the tests to which we subject our theories only test *portions* of those theories, then even highly successful theories may well have

> central terms that are non-referring and central tenets that, because untested, we have no grounds for believing to be approximately true.... In short, to be less than a holist about theory testing is to put at risk precisely that predilection for deep-structure claims which motivates much of the realist enterprise. (1981 116, original emphasis)[2]

Of course, this counterargument succeeds only if defenders of selective confirmation cannot offer any systematic criteria distinguishing the specific parts of a theory confirmed by particular successes from the rest (or if any criterion for so doing cannot include the theory's central theoretical commitments among those portions it picks out as confirmed by the evidence). Psillos attacks this presumption, insisting that because some theoretical claims play an essential role in successful predictions and explanations while others are idle, "it is entirely consistent to stress that empirical evidence sends its support all the way up to the theoretical level, while recognising that it does not do so indiscriminately and without differentiation" (1999 126). Kitcher traces the fundamental insight to Hempel's classic work on confirmation, suggesting "[o]ne obvious moral is that confirmation does not accrue to irrelevant bits of doctrine that are not put to work delivering explanations or predictions" (1993 142n) and concluding that Hempel showed long ago why the sort of indiscriminately holistic theory confirmation Laudan envisions will have to give way to a more differentiated account in any case.[3] And Kitcher and Psillos each go on to discuss particular historical examples in detail in an effort to show that, as a general matter, the successes of past theories did not in fact depend upon their subsequently abandoned claims and presumptions.

But as it stands this appeal to the strategy of selective confirmation faces a crucial problem that appears to be unrecognized by its architects: of any past successful theory the realist asks "What parts of it were true?" and "What parts were responsible for its success?" but *both* questions are answered by appeal to our own *present* theoretical beliefs about the world. That is, one and the same present theory is used *both* as the standard to which components of a past theory must correspond in order to be judged true *and* to decide which of that theory's features or components enabled it to be successful. With this strategy of analysis, an impressive retrospective convergence between our judgments of the sources of a past theory's success and the things it "got right" about the world is virtually guaranteed: it is the very fact that some features of a past theory survive in our present account of nature that leads the realist *both* to regard them as true *and* to believe that they were the sources of the rejected theory's success or effectiveness. So the apparent convergence of truth and the sources of success in past theories is easily explained by the simple fact that both kinds of retrospective judgments have a common source in our present beliefs about nature.

Another way to see the problem here is to imagine the defenders of a later false theory (say Weismann's) answering the realist's questions about an earlier false theory (say Darwin's). Weismann would certainly have agreed that it was the parts of Darwin's theory that it got right (transmission of material units of heredity from parents to offspring, individual hereditary particles responsible for developmental fate of individual cells, etc.) which were responsible for its successes and would certainly have argued that the parts it got wrong (the maturational conception of heredity, gemmules transmitted from somatic tissues to sex organs, etc.) played little or no role in generating the theory's impressively unified explanations of patterns of resemblance and variation between parents and offspring, reversion, bud-variation, graft-hybrids, parthenogenesis, the development of complex tissues, the processes of repair (and their precision), the continuity between various forms of reproduction, and the possibility of producing identical organisms by both budding and seminal generation (with or without complex metamorphoses). But this would only be because Weismann's own theory would have served both as the standard for judging what Darwin got right and as the foundation for understanding how his theory was able to enjoy the successes it did. Notice, for instance, that Weismann's theory would judge Darwin's commitment to the claim that individual hereditary particles are responsible for the developmental fate of particular individual cells to have been both something that Darwin got right and an important source of the theory's successes, despite the fact that current theory judges this to be a fundamental mistake shared by Darwin and Weismann alike. Thus, the historical record is virtually guaranteed to exhibit the pattern that Kitcher and Psillos claim simply because we *interpret* the successes of past theories in the light of our present theoretical beliefs: this pattern will obtain *whether or not* the current theories we are using to assess both the accuracy of parts of past theories and the sources of those theories' success are themselves even approximately true.

None of this would be vicious, of course, if the realist were free to assume that our present theories are in fact true (or approximately true, or probably approximately true, or whatever) but this is just what she cannot assume even provisionally without begging the question against those who are suspicious of realism in the first place. So her point must be more subtle: it must be that the truth of our present theories would explain the substantial match we find between the components of past theories that present theories hold to be true and the components that they hold to be responsible for the success of those theories. But of course the point here is that the impressive magnitude of this correspondence is just as plausibly explained by the common *source* of our judgments about the truth of past theories and the grounds for their success.[4] Thus, as things stand this realist reply to the historical challenge is either question-begging (if it assumes the truth of present theories) or unconvincing (if it simply fastens on one explanation among several plausible alternatives for the substantial correspondence that it finds).[5]

What this virtual guarantee of convergence between our *retrospective* judgments of the sources of success in past theories and the truth (and descriptive accuracy, and referential status) of their parts makes abundantly clear is that any convincing defense of realism by appeal to the strategy of selective confirmation will have to provide us with criteria that *could have been* in the past and *can now be* applied *in advance of any future developments* to identify those idle features or components of scientific theories that are not really confirmed by the empirical successes those theories enjoy. Without such *prospectively applicable* criteria of idleness and/or selective confirmation, the correspondence the realist is able to demonstrate between the parts of past theories that present theories judge to have been true and the parts they judge to have been responsible for the successes of those past theories simply cannot do any real work in defending the realist's position from historical challenges.

The problem is rendered still more acute by the fact that history reveals our prospective judgments in this matter to be demonstrably unreliable. That is, we have already seen enough of the historical record to recognize that the realist will not be able to appeal to the considered and expert judgments of scientists themselves concerning which parts of their theories are really critical to or necessary for their successes without inviting yet another Pyrrhic victory for realism itself. The passages cited from Maxwell, Weismann, and Lavoisier in the previous chapter illustrate that past scientists have repeatedly *misidentified* those parts, features, or aspects of their theories that (by realist lights) were genuinely implicated in or required for their successes: Maxwell was as clear and explicit as he could be in insisting that the wave theory's successful (or even coherent) application required the existence of a material and mechanical ether,[6] Weismann insisted that presumption of germinal specificity was absolutely crucial to his (or any) account's ability to explain the ontogenetic differentiation of cells, and Lavoisier (as we will see in greater detail later on) insisted that the caloric theory's postulation of a subtle fluid was central to the impressive thermodynamic explanations it was able to achieve.

Thus, if the realist is forced to rely on our own contemporary judgments about which parts of our theories are genuinely required for their successes, she will once again concede everything her opponent needs for a convincing historical case against realism: we will perhaps be able to rest confident that the parts of our theories genuinely responsible for their successes (by the lights of later theories) will be preserved in those later theories, but we will be completely unable to depend on our (historically unreliable) judgments about which parts of our own theories those will turn out to be. Thus, just as Psillos's defense of reference for the central theoretical terms in past successful theories left us unable to rely on our ability to discriminate which of a theory's descriptions of an entity were *actually part* of its "core causal description," the strategy of selective confirmation risks leaving us unable to trust our ability to determine, at the time a theory is a going concern, which parts, features, or aspects are *actually required* for the successes of that theory.

Accordingly, without some *prospectively applicable* and *historically reliable* criterion for distinguishing idle and/or genuinely confirmed parts of our theories from others, the strategy of selective confirmation offers no refuge for the scientific realist. Without such a criterion we can have no confidence in our ability to pick out the parts of theories needed for (and thus selectively confirmed by) their successes while those theories are live contenders. We will therefore not be in a position to identify those parts or features of our own theories we may safely regard as accurate descriptions of the natural world (even though we know that not all are), and thus the realist's opponent will again be entitled to the conclusion she wanted all along. And the remainder of this chapter will argue that such criteria are just what existing appeals to selective confirmation do not (and perhaps cannot) provide.[7]

7.2 Theoretical Posits: They Work Hard for the Money

We might begin by considering Kitcher's suggestion that the genuinely confirmed and idle parts of our theories are to be separated by distinguishing their "working posits (the putative referents of terms that occur in problem-solving schemata) from presuppositional posits (those entities that apparently have to exist if the instances of the schemata are to be true)" (1993 149). It is this distinction which grounds Kitcher's further claims against Laudan's reading of the historical record, including the suggestion that "[t]he moral of Laudan's story is not that theoretical positing in general is untrustworthy, but that presuppositional posits are suspect" (1993 149) and that "the success of a theory provides grounds for thinking that . . . the hypotheses that characterize 'working posits' . . . are approximately true" (2001a 170).

Of course, it seems easy enough to reconstruct the predictive and explanatory schemata of past theories so as to avoid appealing to entities abandoned by contemporary theories and then declare them "merely presuppositional," while making prominent use of those theoretical terms and entities that have been retained in current theories.[8] But Kitcher usefully elaborates how merely presuppositional or idle status is to be established with a specific example, arguing that the mechanical ether in which nineteenth-century wave theorists held light waves and other electromagnetic phenomena to be propagated is "a prime example of a presuppositional posit" (1993 149) because the wave theory's successes did not require (and therefore should not have been taken to confirm) its theoretical postulation of any such mechanical medium:

> Modern treatments do not make any reference to an all-pervading ether in which light waves are propagated. But, for Fresnel and many of those who followed him, the existence of such an ether was a presupposition of the successful schemata for treating interference, diffraction, and polarization, apparently forced upon wave theorists by their belief that any wave propagation requires a medium in which the wave propagates. All the successes of the schema can be preserved, even if the belief and the presupposition that it brings in its train are abandoned. That, to a first approximation, is what happened in the subsequent history of wave optics. (1993 145)

On this ground Kitcher replies to Laudan's rhetorically powerful invocation of Maxwell's contention that (in Laudan's words) "the aether was better confirmed than any other theoretical entity in natural philosophy":

> Although we can understand this claim, based as it was on the multiplicity of phenomena to which schemata appealing to wave propagation had been successfully applied, Maxwell was wrong. The entire confirmation of the existence of the ether rested on a series of paths, each sharing a common link. The success of the optical and electromagnetic schemata, employing the mathematical account of wave propagation begun by Fresnel and extended by his successors (including, of course, Maxwell), gave scientists good reason for believing that electromagnetic waves were propagated according to Maxwell's equations. From that conclusion they could derive the existence of the ether—but only by supposing in every case that wave propagation requires a medium. Thus the confirmation of the existence of the ether was no better than the evidence for that supposition. (1993 149)

If any prospectively applicable criterion of selective confirmation is on offer here at all, it seems we must understand Kitcher to be suggesting that the ether was an idle or merely presuppositional rather than a working posit simply because there was an alternative possible account on which the empirical predictions and explanatory schemata of the wave theory could survive while the belief in a material ether was simply erased from the theory: to wit, that light and/or electromagnetic disturbances are transmitted *like* mechanical waves (and according to Maxwell's equations), but (somehow) without any medium of transmission at all. But this criterion of eliminability or idleness is far too easily satisfied to do the work that Kitcher requires of it, for it applies perfectly straightforwardly to virtually all those posits of contemporary theories that realists hope to defend as *genuinely* confirmed by their successes, including Kitcher's own paradigmatic examples of "working" posits: the electromagnetic field, genes, atoms, and molecules. We can just as easily suppose, for example, that characteristics are passed from parent to offspring in the patterns suggested by contemporary genetics, but (somehow) without the existence of genes themselves. The existence of genes *seems* to be confirmed by the role they play in innumerable explanations of the inheritance and production of the characteristics of organisms, but only by supposing in each case that the inheritance and production of phenotypic traits requires that there be some material causal basis for those characteristics that is transmitted between parents and offspring. Thus, if the existence of such a spare alternative as "transmitted like waves, but without any medium of transmission" suffices to render the ether "idle" or "presuppositional" (and therefore unconfirmed by the successes of the wave theory), then virtually *any* theoretical posit must be so regarded.

In fact, the problem for Kitcher here runs far deeper, for we have already seen that Maxwell himself explicitly considered whether the wave theory's successes could survive the excision of the ether from the theory and rejected the *very intelligibility* of the resulting idea of wave transmission without any mechanical medium. Consider once again the closing words of *A Treatise on Electricity and Magnetism* (quoted in the previous chapter):

> If something is transmitted from one particle to another at a distance, what is the condition after it has left the one particle and before it has reached the other? If this something is the potential energy of two particles, as in Neumann's theory, how are we to conceive this energy as existing in a point of space, coinciding neither with the one particle nor with the other? *In fact, whenever energy is transmitted from one body to another in time, there must be a medium or substance in which the energy exists after it leaves one body and before it reaches the other,* for energy, as Torricelli remarked, 'is a quintessence of so subtile a nature that *it cannot be contained in any vessel except the inmost substance of material things*'. Hence all these theories lead to the conception of a medium in which propagation takes place. . . . (1955 [1873] Vol. II 493, my emphasis)

This passage makes clear that Maxwell explicitly considered the alternative whose existence is supposed to render the ether an idle or merely presuppositional posit—that electromagnetic phenomena are transmitted in a wavelike fashion but without any mechanical medium of transmission—*and rejected this very idea as incoherent.*[9] But if the alternative that eschews appeal to a given theoretical posit and thereby undermines its confirmation by the successes of the theory that posits it need not even be *intelligible* to the scientists who hold that theory, then the realist seems to lose any hope whatsoever of insulating any theoretical posits of current science from idleness or merely presuppositional status. Perhaps, for instance, material objects are made up of smaller, atomic constituents, but (somehow) without the existence of those atomic constituents themselves: if Kitcher's discussion of the ether is our model, our present judgment that this suggestion is incoherent simply poses no obstacle to regarding atoms as idle, eliminable, merely presuppositional, and therefore not genuinely confirmed by the successes of the atomic theory of matter.

Indeed, the electromagnetic field offers a fairly precise confirmational analogue of the ether from which (in Maxwell's and Lorentz's conceptions) it is descended: contemporary theorizing holds that the field in which electromagnetic disturbances are propagated is an entity ontologically on a par with matter itself—that is what seems to *us* to be required if electromagnetism is to be propagated *without* any mechanical and material medium of the sort Maxwell envisioned.[10] But this is surely presuppositional in precisely the same sense that the ether was: electromagnetism could be *somehow* otherwise propagated according to the familiar equations without requiring even the existence of an electromagnetic field itself on an ontological par with matter (especially if intelligibility is no more supposed to be a constraint on our recognition of possibilities than Kitcher's treatment suggests it should have been for Maxwell); thus, *our* predictive and explanatory schemata no more *need* to appeal to the existence of such a field in deriving electromagnetic phenomena than those of the ether theorists were forced to appeal to the existence of an ether. And thus we seem once again to lose the confirmation for the very theoretical entities the realist hopes to defend.[11]

Can we improve on Kitcher's account of idle or merely presuppositional status? Perhaps the ether is merely presuppositional not because it is eliminable in Kitcher's indefensibly weak sense, but because the theory also ascribes no direct causal role to it in the production of optical and electomagnetic phenomena. After all, a nineteenth century physicist asked to explain the occurrence of some particular electromagnetic or optical phenomenon would not have been likely to cite "the ether" or some state of the ether as its cause, even if reference to the role of the ether would certainly have shown up in any more comprehensive causal story she would have been inclined to tell. Although perfectly natural, this suggestion seems to run afoul of any number of discarded theoretical posits that *were* ascribed direct causal roles in the production of phenomena by the successful explanatory and predictive practices of their respective theories. To such familiar examples as phlogiston and caloric fluid we've seen that we might fairly add Darwin's gemmules, Galton's stirp, and Weismann's biophors. Nor does there seem to be any further promising sense in which these entities were merely presuppositional rather than working posits or were disconnected from the explanatory and even predictive successes of their respective theories: these simply *were* the entities most directly responsible for producing systematically related characteristics in various organisms and for other hereditary phenomena. Thus, it seems that causal role considerations will not help Kitcher, for the successful explanatory and predictive practices of past theories have routinely ascribed direct causal roles to subsequently abandoned theoretical entities.

Of course, none of this controverts Kitcher's Hempelian suggestion that any successful account of confirmation will have to refuse confirmation to unconnected parts of a successful theory that are not "put to work" in delivering its predictions and explanations. At the heart of Kitcher's proposal is the idea that the challenge posed by the ether, biophors, and the rest is simply the "tacking" problem we discussed in chapter 1, and as we noted there, that problem is itself a serious one: we cannot allow that the evidence supporting the modern synthetic theory of evolution provides equally strong support to the theory created by "tacking on" to it an unrelated claim like "and jellybeans grow on trees on the planets orbiting Alpha Centauri." And there are certainly genuine scientific cases of claims or commitments that have similarly played no real role in the theories that included them, like Newton's commitment to the universe being at rest in absolute space (cf. chap. 1). What our historical cases suggest, however, is that the rejected posits of past theories, like ether, phlogiston, gemmules, stirps, and biophors (as well as caloric fluid, see below), were simply not any less intimately involved in the predictive and explanatory accomplishments of those theories than genes, atoms, molecules, and the electromagnetic field are in our own. These entities were eliminable from the theories that posited them only in the same trivial sense in which any posit of any current theory could similarly be eliminated, and to convict them of such "jellybean tree" idleness requires a standard so weak as to demand a similar conviction for all theoretical commitments or beliefs whatsoever. Thus, we have no reason to think that whatever solution to the tacking problem we ultimately embrace will prevent confirmation from having accrued to the ether, phlogiston, caloric, biophors, and other central posits of actual past scientific theories, or indeed that any successful account of confirmation can afford to deny that such entities were genuinely confirmed by the accomplishments of the theories that posited them.

7.3 Trust and Betrayal

Stathis Psillos's intriguing alternative approach promises to evade these persistent problems by avoiding the need for any *explicit* criterion of selective confirmation at all. His general appeals to the "essential" and "idle" components of past theories (1999 110f) and his explanations of why specific features of particular theories were inessential for their successes face the same challenges about prospective application we raised above. Yet, he seeks to finesse this problem by arguing that working scientists *themselves* routinely judge different parts, features, or aspects of extant theories to be differentially confirmed by the empirical evidence and that the historical record shows these judgments to be generally reliable. If so, we can defend realism from historical challenges without any explicit account of selective confirmation or any explicit criterion of idleness, for we could have in the past and can now safely rely on *scientists' own* judgments in identifying the selectively confirmed, trustworthy aspects of existing theories, *however* they manage to reach such judgments.[12] Sadly, Psillos's defense of this creative strategy itself turns out to depend upon a highly selective reading of the historical record.

We have, of course, already seen enough historical evidence to be suspicious of Psillos's central claim. To note just a few examples, it was nineteenth-century physiologists and embryologists themselves who argued that vital forces were absolutely required by the distinctively teleological or directive character of organic causal processes (chap. 3), it was Weismann himself who insisted that ontogenetic differentiation could not possibly be explained without germinal specificity (chap. 5), and it was nineteenth-century physicists themselves who judged the ether as well confirmed by the successes of the wave theory as it is possible for theoretical entities to be (see above and chap. 6). Maxwell is fairly described as a leading scientific intellect and was arguably the foremost expert at the time on the wave theory of electromagnetism. For him to misjudge (by realist lights) the confirmational status of the ether in so spectacular a fashion casts considerable doubt on the idea that scientists' own judgments of selective confirmation have been historically reliable.[13]

Psillos does not address any of the examples from eighteenth- and nineteenth-century biological science we have considered, but his discussion of the ether quite rightly notes the agnosticism of many ether theorists regarding the exact constitution of the ether as well as caution and even outright skepticism regarding the details of such particular mechanical models of the ether as Green's elastic solid, McCullagh's rotational ether, and Stokes's elastic jelly. But of course this proves far too little for the realist's needs, for this skepticism regarding particular mechanical models of the ether was more than matched by the same theorists' confidence in light of the available evidence that there must be *some mechanical medium or other* in which light waves were propagated: Green's claim, for instance, that we are "perfectly ignorant of the mode of action of the elements of the luminous ether on each other" (cited in Psillos 1999 132) expresses cautious agnosticism regarding the existing models of the ether's mechanical interaction, but *not* regarding the existence of the ether itself. Similarly, Psillos points out (1999 138) the difference in Maxwell's attitude toward his accounts of the dynamics of electromagnetic propagation in the ether and of the constitution of the ether itself, but that is again not enough. Although Maxwell was (rightly) not at all confident that he had successfully identified the correct mechanical account of the medium in which electromagnetic waves were propagated, we have seen that he was nonetheless as confident as he could be that the evidence supported (indeed necessitated) the existence of *some such mechanical medium or other*. Indeed, G. F. Fitzgerald, writing no later than 1888, uses this very distinction to make the same point in the course of defending the view that electric and magnetic phenomena generally are transmitted through the same mechanical medium in which light waves are propagated. He first argues that "if [electric and magnetic actions] can be shown to be explicable by a medium possessing the same properties as are required in order to explain the transmission of light... these actions will then be explained as due to a known cause." He then continues:

> Before describing, in general terms, the sort of way in which it has been shown that electric and magnetic phenomena may be due to a medium, it may be well to call our attention to how very little we know of the constitution of the media we use every day for transmitting force.... In the case of a clock or watch spring nobody knows what is the exact structure of its molecules by means of which it is able to act as a reservoir of energy, which we put into it when we wind up the clock or watch, and which is gradually expended in keeping the works going. The fact that we do not know this does not in the least diminish from the importance of our knowing that there is a spring, and that the action is not a pure action at a distance between the axle that we turn with the key and the barrel that turns the wheels. Similarly it is of importance to know where in the medium the energy of an electrified system is stored, although we may not be able to state what the exact structure of the medium must be in order that it may be capable of acting as a reservoir of energy. (1902 164–165)

Thus, Psillos's claim (1999 140) that "[t]he parts of 'luminiferous ether' theories which were taken by scientists to be well supported by the evidence and to contribute to well-founded explanations of the phenomena were retained in subsequent theories" can only be defended by conveniently ignoring many such judgments made by leading scientists of the time, including the very scientists to whose more cautious and skeptical attitudes he appeals.

The same sort of selective attention is at work in the one other case Psillos considers in detail: the material fluid or caloric theory of heat. In this connection he effectively documents Joseph Black's scrupulous refusal to regard his experiments as conclusively establishing a material fluid theory of heat against the competing "dynamic" account of heat as motion. What Psillos misses, however, is that this restraint simply reflects Black's unusually strict but characteristically eighteenth-century Scottish commitment to Newtonian inductivism. That is, Black advocates an official hostility toward *all* theories and theorizing *in general*, famously expressed in his warning to his student and editor John Robison to "[reject] even without examination, every hypothetical explanation, as a mere waste of time and ingenuity" (1803 vii). Black ultimately judged Cleghorn's material fluid account of heat "the most probable of any that I know," but immediately reasserted his officially antitheoretical stance, reminding us that "it is, however, altogether a supposition" (1803 33) and insisting later that all such hypothetical suppositions were best avoided "as taking up time which may be better employed in learning more of the general laws of chemical operations" (1803 192–193).[14] Even more troubling for Psillos's thesis is the confidence with which Black *rejected* the dynamical theory in light of the evidence: he argued that his own discoveries concerning latent heat simply could not be squared with the dynamical account (McKie and Heathcote 1935 28), insisting as well that he could not conceive of an "internal tremor" of the particles of substances "that has any tendency to explain, even the more simple effects of heat" and that the dynamic theory's supporters "have been contented with very slight resemblances indeed, between those most simple effects of heat, and the legitimate consequences of a tremulous motion" (1803 31). Psillos does not, then, discover any *special* or *selective* reticence on Black's part to endorse the material fluid theory of heat in light of the evidence, and insofar as Black made any judgments of *selective* confirmation at all, they are ones that realists must regard as profoundly mistaken: that the material fluid account was the most probable theory of heat and that his own discoveries definitively refuted the competing dynamical view.

Black's general hostility toward theories was certainly not shared by Lavoisier, who advocated instead a "systematic" approach to chemistry and argued against simply piling up empirical facts (see Guerlac 1976 219); accordingly, the most impressive piece of textual evidence Psillos cites in support of his historical thesis is Lavoisier and Laplace's explicit insistence in opening their famous *Mémoire sur la Chaleur* that the experiments therein are consistent with either a material fluid or a dynamical theory of heat and that they will not decide between the two. Upon further investigation, however, this evidence carries less weight than it seems to, for the point of the *Mémoire* is simply to present Lavoisier and Laplace's new ice calorimeter and its techniques for the measurement of heat and the specific heats of particular substances: "We have already said, and we cannot stress this fact too much, that it is less the result of our experiments than the method we have used that we offer to scientists..." (Lavoisier and Laplace 1982 [1783] 32). Since these new calorimetric methods really were compatible with both the material and dynamical theories of heat, it is unsurprising that Lavoisier and Laplace address their new techniques to the widest possible audience of their interested contemporaries.[15] Furthermore, the explanations of specific phenomena offered later in the joint *Mémoire* itself are indeed committed to the view that heat is a material substance (see Morris 1972 30–31), most notably in their appeal to the chemical *combination* of caloric with other substances (e.g., in phase transitions).

Far more important than Lavoisier's reasons for wanting to present himself cautiously in (at least the first half of) the *Mémoire sur la Chaleur*, however, are his repeated endorsements of, explanatory appeals to, and confident judgments of confirmation for the material fluid theory itself in his many other works. Throughout his career Lavoisier steadfastly defended a material fluid theory of heat in which the "matter of fire" could exist in either free or combined forms and whose central contention was that the state (solid, liquid, aeriform fluid) and changes of state of any matter depended upon its *chemical combination* with this matter of fire, an account of the matter that cannot be straightforwardly transformed into dynamical terms.[16] As early as his 1778 paper on the formation of elastic fluids Lavoisier claimed not only that the opinion that there is such a material fluid was "that of the great majority of ancient physicists"[17] and that he could therefore "dispense with reporting the facts upon which it is founded," but also that "the collection of memoirs . . . that I have to offer will serve to prove it," in part by showing "that it agrees everywhere with the phenomena, that it everywhere explains everything that happens in physical and chemical experiments" (*Oeuvres de Lavoisier* (hereafter *OL*) v. II 212, my translation). Lavoisier appealed to this account alone throughout his works to explain a wide range of phenomena, including latent heat, the specific heats of substances, evaporative cooling, the elasticity of gases, and (along with his "oxygen principle") the phenomena of combustion and calcination. Indeed, the material fluid theory of heat plays a critical role in his case against phlogiston and for the new chemistry: Lavoisier's discussion of an alternative to the phlogiston theory in his famous antiphlogistic memoir of 1785 consists almost entirely of presenting his material fluid theory of heat and showing how it can explain a wide range of thermal phenomena (including the relative quantities of heat released by various combustions and calcinations, the heat produced by mixing water and concentrated vitriolic acid, and the heat lost when salt is dissolved in water) and can make general predictions of whether heat will be gained or lost in a given reaction (see Morris 1972 17f). Perhaps most important of all, the material theory of heat provided Lavoisier with a viable alternative explanation of the release of fire during combustion, the primary explanatory accomplishment of the phlogiston theory (see Guerlac 1976 202–203, Morris 1972 37–38).

The memoir on phlogiston does mark Lavoisier's increasing concern to integrate his earlier, exclusively chemical, discussion of thermal phenomena (in terms of free and chemically combined caloric fluid) with a physical account of these phenomena in terms of interparticulate forces and the availability of the spaces between physical particles to receive caloric fluid. But this shift does not strengthen Psillos's case, as Lavoisier's increasingly physical presentations of the account remain forcefully committed to the material fluid view:

> It is clear *a priori* and *independent of all hypotheses* that the greater the distance between the molecules of bodies, the greater their capacity to receive the matter of heat, and consequently their specific heat will be

> greater. (Lavoisier 1785 as translated in Donovan 1993 172, last emphasis mine)

Most important of all, neither Lavoisier's endorsement of the account nor his judgment that it is well confirmed by the existing evidence is at all reserved or qualified in the way Psillos suggests. Remarking on the expansion and contraction of bodies, Lavoisier says:

> One can hardly think about these phenomena without admitting the existence of a special fluid [whose accumulation causes heat and whose absence causes cold]. It is no doubt this fluid, which gets between the particles of bodies, separates them, and occupies the spaces between them. Like a great many physicists I call this fluid, whatever it is, *the igneous fluid, the matter of heat and fire.* (Lavoisier 1785 as translated in Donovan 1993 171, original emphasis, translation modified; quoted in chap. 6)

Nor is Lavoisier's enthusiasm for the material theory (unlike his calculated expression of agnosticism in the *Mémoire sur la Chaleur*) confined to just a single work or a single moment in his career: both the *Traité de Chimie* (vol. I. 1–3) and the *Mémoires de Chimie* (vol. I. 3–5) offer accounts of the confirmational significance of expansion and contraction for the material fluid account that are virtually identical to that given in the antiphlogistic memoir of 1785. The *Traité*, for example, claims that:

> It is difficult to conceive of these phenomena without admitting that they are the result of a real, material substance, of a very subtile fluid, that insinuates itself throughout the molecules of all bodies and pushes them apart. . . . This substance, whatever it is, is the cause of heat, or in other words, the sensation that we call heat is the effect of the accumulation of this substance. . . . (*Traité de Chimie* vol. I 1–3 in *OL*, my translation; quoted in chap. 6)

As Psillos notes (see also Kitcher 1993 278n), the *Traité* does contain Lavoisier's one other explicit note of caution recognizing the hypothetical status of the material fluid: "strictly speaking," Lavoisier says, it is sufficient "that it is considered as the repulsive cause, whatever that may be, which separates the particles of matter from each other" (*Traité de Chimie* vol. I 5, cited in Psillos 1999 119). Once again, however, Lavoisier's later explanations of specific phenomena in the same work "take for granted the existence of heat matter as a fact of nature not requiring justification" (Morris 1972 31). More importantly, this recognition of the hypothetical status of the igneous fluid did not *generally* lead Lavoisier to qualify his endorsement of the material fluid theory of heat or his confidence in its confirmation by the existing evidence, leading Morris (1972 30–31) to the blunt conclusions that "Lavoisier never faltered in his belief that the cause of sensible heat is a material substance" and that his "true feelings regarding this 'singular hypothesis'" are those expressed in his posthumously published final collection of essays, the *Mémoires de Chimie*. The discussion in that work is perhaps the most damaging of all for Psillos, as Lavoisier there directly and explicitly confronts the suggestion that the "igneous fluid" is merely hypothetical, and responds with a forceful and explicit defense of its conclusive confirmation by the existing evidence:

> once I have shown, in the collection of memoirs I am publishing, that it is everywhere in accordance with the phenomena, that everywhere it explains in a simple and natural manner the result of experiments, this hypothesis will cease to be one, and one will be able to regard it as a truth. (*Mémoires de Chimie* vol. I 2, in *OL*, my translation)

The strength of Lavoisier's confidence in the confirmation of the material theory of heat by the available evidence is all the more remarkable in view of the caution and diffidence with which he is famous for having typically presented claims that he regarded as theoretical or hypothetical (see Donovan 1993 168). Thus, I do not see how the textual evidence can be reconciled with Psillos's claim that "scientists of this period were not committed to the truth of the hypothesis that the cause of heat was a material substance" (1999 119). And once again, it seems that Psillos's case for the reliability (by current lights) of scientists' own judgments of selective confirmation must itself rely on an extremely selective reading of the historical record that ignores or dismisses many if not most of those very judgments.

Finally, as these historical cases illustrate, Psillos's account also seems to require an implausible *homogeneity* in scientists' own judgments of selective confirmation: the assurance that we can trust scientists' own judgments will not help us decide what to believe in the routine case of *disagreement* among contemporary scientists concerning which parts of extant theories are well confirmed by the evidence—and this would be true even if we ignored the fact that such disagreements were a ubiquitous feature of past science. And if Psillos means to claim only that we can rely on scientists' own judgments of selective confirmation when they are in fact univocal, then the realism he seeks to defend would be slender indeed, even if his claim of reliability for scientists' own judgments of selective confirmation were not historically indefensible.

Of course Psillos might suggest that judgments of selective confirmation are complex and difficult and that Maxwell, Lavoisier, Weismann, and others have simply gotten them wrong. But this escape concedes the pointlessness of the detour through the historical record of (unreliable, as it turns out) judgments of selective confirmation by scientists themselves and thus inherits again the very problem that Kitcher could not solve and that Psillos sought to finesse: providing an explicit, historically reliable criterion for identifying the parts of theories selectively confirmed by their successes that *prospectively* distinguishes the rejected posits of past theories from those of the present theories that the realist hopes to defend. It seems, then, that the unfulfilled demand for prospectively applicable criteria of idleness will prevent the strategy of selective confirmation from protecting the realist from the weight of the historical evidence.

7.4 Structural Realism and Retention

Once the general character of this unsatisfied demand for historically reliable and prospectively applicable criteria is recognized, it can be seen to undermine even more modest attempts to defend realist inferential entitlements as well. Perhaps the most influential and promising form of such a modest realism is the position that John Worrall has ably defended under the name "structural realism."

To his credit, Worrall (1989, 1994) keenly feels the pull of both the explanationist defense of realism and the pessimistic induction, but he thinks that a close look at the details of the historical examples creating trouble for the realist shows us how we might do honor to both arguments. The example that impresses him most is one we have considered in some detail—the transition from Fresnel's wave theory of light to the later electromagnetic theories that replaced the postulation of a mechanical, material luminiferous ether with that of the electromagnetic field. While Worrall concedes that Fresnel's theoretical account cannot in any reasonable sense be said to be approximately true, what he finds most striking about the example is that Fresnel's mathematical description of the propagation of light waves is simply carried forward intact and unchanged from his own theory to its successor: thus, *something* critically important survives the replacement of Fresnel's wave optics by the later theory of the electromagnetic field, and with some justice Worrall is inclined to characterize this as the theory's claims about the *structure* of the natural phenomena it describes. Following some remarks of Poincaré, Worrall uses this particular case to motivate the suggestion that, as a quite general matter, a predictively successful theory's claims about the structure of natural phenomena survive and are preserved in its successors (at least as limiting cases, see 1989 120), even as what he variously characterizes as those same theories' claims about the "nature" of the entities they describe (1989 117–118, 1994 334), their "content" (1989 117, 1994 340), or their "ontology" (1994 336, 341) are abandoned and/or replaced wholesale. It is therefore only the structural claims of our predictively successful theories, he suggests, that are the legitimate objects of justified realist confidence. A successful defense of this "structural realism" would at least qualify the impact of the problem of unconceived alternatives (along with that of the original pessimistic induction) in a now-familiar fashion, for it would suggest that we can be confident in the claims of our successful current theories about the structure of nature, even if there are indeed serious and well-confirmed alternatives to those theories remaining presently unconceived.

But the problem with Worrall's suggestion will by this point also have a familiar ring: it is not at all clear that we can plausibly distinguish the claims of a theory about the *structure* of natural phenomena from its "content," "ontology," or claims about the "nature" of the entities it describes even in the case of current theories, much less assess the record of past practitioners in making such discriminations successfully. But (to invoke the now-familiar refrain), if the structuralist defense is to be of real service to realism, it must either offer a prospectively applicable criterion for identifying the structural features of theories that is also historically reliable (i.e., would have picked out just those features of past theories that would be preserved in their successors), or it must show that our own ability to make this same discrimination successfully without any explicit criterion has been historically reliable. Unless we can do one or the other, this defense of structural continuity between predictively successful scientific theories will simply not put us in a position to know which claims or features of our own theories are those on which we may rely.

Appeals to vague intuitions simply will not do here: at best such an intuitive criterion renders the problem of reliable prospective application especially acute, forcing the structural realist to qualify her beliefs in even the structural claims of a theory by her level of confidence that they are indeed its structural claims. But even worse, a merely intuitive criterion of structure seems to run afoul of the historical record in what seem central and obvious cases. Weismann famously made use of his early account of inheritance, for example, to predict the existence of reduction division (chapter 5). But what prevents the account's (mistaken and subsequently abandoned) insistence that the germinal materials passed from parents to offspring must be parceled out differently to heterogeneous parts of an organism's body from being a claim about the *structure* of the processes underlying inheritance, ontogeny, and/or hereditary phenomena?

Perhaps in recognition of the sort of difficulties invited by an intuitive criterion, Worrall sometimes offers the more concrete suggestion that the structural commitments of a theory consist simply of its equations or the abstract mathematical relationships it posits (1989 118–120, 1994 340). But there would seem to be something extremely misleading in saying even that the abstract mathematical relationships posited by past successful theories have described the "structure" of the natural world in ways that are still embraced by current theories. To take just one further example, the most widely influential and most predictively successful (see below) aspect of Galton's stirp theory of inheritance (chapter 4) was its central mathematical formalism: the "Ancestral Law of Inheritance," which specified the proportion of hereditary material in the stirp of any organism contributed by each of its ancestors. According to this law:

> the two parents contribute between them on the average one-half, or (0.5) of the total heritage of the offspring; the four grandparents, one-quarter or $(0.5)^2$; the eight great-grandparents, one eighth or $(0.5)^3$, and so on. Thus the sum of the ancestral contributions is expressed by the series $\{(0.5) + (0.5)^2 + (0.5)^3$, &c.$\}$, which, being equal to 1, accounts for the whole heritage. (Galton 1897 402)

Of course, it is true enough to say that the fractional relationships described by Galton's Ancestral Law show up *somewhere* in the account of inheritance offered by contemporary genetics. As William Provine points out, "[b]y 1900, when Mendelian heredity was rediscovered, Karl Pearson had moved the law of ancestral heredity to a purely phenotypic level independent of whatever physiological mechanism of heredity might be operating" (1971 181). And Robert Olby notes that "[t]oday Galton's Ancestral Law of Inheritance still stands as a mathematical representation of the average distribution of continuously varying characters in a population of freely outbreeding individuals not subject to selection" and it "serves as a basis for predicting the average distribution of such characters in the population" (1966 81–82; see also 1987 403, 409).[18] Thus the formal mathematical relationship described by the Ancestral Law can certainly be unearthed by sufficiently persistent digging into the corners of the theoretical description of the world given to us by contemporary genetics.

But it is equally true that contemporary genetics does not recognize the fractional relationships expressed in Galton's Ancestral Law as describing any fundamental or even particularly significant aspect of the mathematical structure of inheritance. This seems especially clear if we keep in mind that the Ancestral Law is expressed above in terms of *generational* contributions to the stirp of the offspring: as Galton is careful to point out, each individual parent contributes only 1/4 of the makeup of an organism's stirp as a whole (for a total parental contribution of 1/2), each individual grandparent contributes only 1/16 (for a total grandparental contribution of 1/4), and so on.[19] And contemporary genetics simply does not recognize anything in a given organism as inherited from or corresponding to its ancestors in this fractional distribution.[20] Thus, to evade the challenge of the historical record the mathematized version of structural realism will have to retreat simply to the dogged insistence that such chunks of mathematical formalism as Avogadro's number, Fresnel's equations for the transmission of light, and Galton's Ancestral Law of Inheritance will be recoverable in *some* way, *some*where, *some*how from future science. This, of course, is a far cry from giving us *any* claim we can rely on as an accurate description of (even just the structure of) the natural world, and therefore invites the trust argument again with a vengeance. Like the others we've examined, then, Worrall's structuralist defense of even a restricted form of realism seems to give the game away entirely: it either leaves us with no justifiable confidence in our ability to clearly distinguish those claims of current theories about the natural world on which we may rely, or it forces us to draw such a distinction in a way that invites yet another renewed pessimistic induction over the historical record itself.

7.5 Selective Confirmation: No Refuge for Realism

In the last two chapters we have examined the most influential recent efforts to respond to the pessimistic induction and to recruit the historical record to the defense of scientific realism. We have done so because a number of these responses, if successful, seemed to promise relief from the problem of unconceived alternatives as well as the pessimistic induction itself: they promised to defend the realist's confidence in the truth of (at least some parts of) our best contemporary theories, *even if there are* indeed well-confirmed and serious scientific alternatives to them that remain presently unconceived. What we have found, however, is that despite considerable ingenuity and sophistication these responses exemplify a curiously persistent pattern of systematic argumentative failure. The simplest replies manage to defend the letter of some original realist claim (for example, that the central terms in successful theories have referred, or that successful theories have turned out to be approximately and/or partially true), but only in senses that run them immediately afoul of the trust argument: they leave us free to concede the claims of the proffered defense but still answer "no" to the crucial question at the heart of the matter—whether we should *trust* or *believe* the accounts of otherwise inaccessible natural domains given by our best scientific theories. Other replies, recognizing this result as self-defeating for the realist, have sought instead to distinguish some parts of past theories from others and defend realist inferential entitlements only concerning those parts of successful scientific theories that exhibit some distinctive further characteristic, such as the "core causal descriptions" associated with the theory's central terms, the theory's "working" rather than "presuppositional" or "idle" posits, or the theory's claims about the "structure" of the phenomena. But these more sophisticated defenses have turned out to provide similarly Pyrrhic victories for the realist. The criteria they offer us for distinguishing the parts of present theories that will be preserved or retained in their successors are either not prospectively applicable at all, fail to distinguish the parts of present theories realists hoped to defend from any others, or require us to exercise discriminatory abilities whose reliability is itself subject to an historical challenge as compelling as any that we faced from the original pessimistic induction. Thus, none of these realist defenses can offer us any justifiable confidence in our ability to pick out the parts, aspects, or features of our best contemporary scientific theories on which we may rely. The upshot is that the most influential and sophisticated recent defenses of realism from the historical record do not constitute an effective reply to the original pessimistic induction to which they were addressed, much less give us any reason to qualify the significance we attach to the problem of unconceived alternatives. We have not shown that no possible version of any of these strategies can be made to work, but the reliability with which realist offerings seem to run into some version of this same tangle of difficulties is a discouraging sign.

It is important, however, that we draw the right moral from these failings of the realist reply to history and from what seems to be the systematic vulnerability of theoretical natural science to the problem of unconceived alternatives. The moral is not that we can simply never trust the deliverances of our scientific investigations of the world—nothing so grand (or ridiculous) has been defended or even suggested here. Instead, what we've seen is that the challenge to our scientific beliefs posed by the problem of unconceived alternatives is quite genuine, and that the most promising proposals concerning how we might circumscribe the problem so as to insulate all or even some systematically identifiable set of scientific claims from the challenge simply will not work. Just as important, we've seen that we cannot trust the brute intuitions of either laypersons or professional scientists concerning whether or when a given scientific theory's success or confirmation by the existing evidence is sufficient to warrant the judgment that it or any particular part of it is true. Thus in matters of convincing confirmation, we cannot responsibly rest content with Justice Stewart's dictum[21] regarding pornography: that he knew it when he saw it. Nor can we responsibly embrace the reply that the evidence itself leaves us more convinced of the truth of our scientific theories (or a particular scientific theory) than of the seriousness of the problem of unconceived alternatives itself, notwithstanding the current fashion for this sort of argument in professional philosophical discourse. If this is much less than a wholesale rejection of the truth of claims grounded in our best scientific theories, it is much more than a simple methodological corrective or supplement to current scientific orthodoxy, for until we determine the reach of the problem of unconceived alternatives, we will be quite unsure of what we can justifiably trust or rely on in our scientific description of the natural world.

The daunting task that remains, then, is that of figuring out exactly how, why, where, and when the problem matters in particular cases. That is, we must begin to ask exactly where and when our scientific beliefs rest on eliminative foundations that are vulnerable to the problem, and what epistemic attitude should replace simple belief for the scientific claims that turn out to be vulnerable in this way, as certainly appears to be the case for our most general and fundamental scientific theories. The first of these tasks will be as wide-ranging and involved as theoretical science itself. In the final chapter we will make a beginning on the second, though only a beginning, by first asking whether there is any coherent attitude toward the claims of theoretical science that remains open to us if scientific realism is abandoned.

Notes

1. Leplin (1997 146–151), for example, endorses this strategy of response to the pessimistic induction but does not develop or argue for it himself, contenting himself instead with challenging opponents of realism to show that the historical record does *not* exhibit the pattern Kitcher (see below) has claimed.

2. For a more extensive development of realist's difficulties on this score, in connection with her accounts of theoretical equivalence and explanation, see Sklar (1982).

3. The point of Hempel's original line of thought is easily seen in a simple example. It cannot be the case that any true implication of a theory automatically confirms all the statements making up that theory. If this were so, we could generate spurious confirmation for any statement at all! To use an example familiar from chapter 1, we could simply create a new "theory" by adding to (say) contemporary molecular genetics the further claim "and jellybeans grow on trees on the planets orbiting Alpha Centauri." This new theory shares all the true implications of molecular genetics itself, so if these confirm every claim of the theory they confirm the claim about the distant jellybean trees orbiting Alpha Centauri, too. As the familiar example suggests, this just is a version of the "tacking" problem discussed earlier, and it has convinced many philosophers that a purely hypothetico-deductive approach to confirmation is simply unworkable.

4. Note that this problem is not obviated by the existence of some cases in which even the realist admits that a given empirical success enjoyed by a rejected theory relied upon false parts of that theory, for interpreting the fortunes of past theories in terms of present accounts of nature cannot ensure that some correspondence to a present account will be available to explain *every* success of a rejected theory. It does ensure, however, that the convergence or correspondence we *do* find between the true parts of past theories and the sources of their success is as well explained by our use of a common source to judge truth and sources of success in past theories as it would be by the truth of both present theories and the corresponding parts of past ones.

5. This same challenge applies to Richard Boyd's influential (1990) defense of realism, for he assumes (as he recognizes) the truth of present accounts of nature in defending such distinctive realist theses as the maturity of recent scientific theories, the increasing reliability of scientific methods, and the "retrospectively sustained mutual ratification" of theoretical developments in the sciences. It is perhaps for this reason that Boyd suggests that realism comes as a "package" that must be weighed against competing empiricist or constructivist packages.

6. This example is especially significant, as it shows that we have in central cases misidentified even those parts of our theories responsible or required for their successes *in novel prediction*. It is perhaps worth noting explicitly here that Maxwell was far from alone in judging the successes of the wave theory to absolutely require the postulation of a mechanical and material medium of transmission. Consider, for example, the following remark from G. F. Fitzgerald (written no later than 1888), describing an earlier state of research in which light was known to propagate with a finite velocity (and thus judged to require a mechanical medium) while electric interactions were taken to involve action at a distance: "It is only when you wish to take account of actions that take some time to be transferred from you to the body, it is only then that it is necessary mathematically to introduce symbols expressly referring to a medium. We may gather from this how it was absolutely necessary in the case of the action of light to introduce the notion of something existing between the Sun, for instance, and the Earth, while it was not necessary, as far as was known, to make a similar assumption in regard of electrical actions" (Fitzgerald 1902 163–164).

7. I will also argue in section 7.4 that John Worrall's Structural Realism suffers from a related infirmity, in that those "structural" elements of successful theories that Worrall claims are typically preserved in their historical successors seem *identifiable* as distinctively structural only in retrospect.

8. Indeed, Psillos himself makes this point forcefully (1999 111–112) in criticizing Kitcher's backward-looking distinction between presuppositional and working posits as inadequate.

9. This would seem to be what in turn grounded his confidence that the ether's material and mechanical character must be part of its "core causal description"—the description that anything would have to satisfy in order to play the causal role ascribed to the ether—the point this same passage was used to make in the previous chapter.

10. In fact, contemporary quantum field theory gives us some reason to wonder whether there is any real distinction between matter and states or configurations of a field, but we may safely neglect this complication for present purposes.

11. As this example also illustrates, recognizing either such spare and/or apparently incoherent alternatives as legitimate alternatives capable of undermining the confirmation of allegedly idle or presuppositional posits also exposes the realist to an especially pointed version of the problem of underdetermination (cf. Worrall 1994).

12. Note that this is quite a different proposal than the one we considered at the outset of this chapter: there we asked if scientists' own judgments of which parts of their theories were *responsible for their successes* have historically proved to be reliable (and determined that they have not). On Psillos's strategy, we must ask instead whether scientists' own judgments concerning which parts of their theories are *differentially confirmed* by the evidence have historically proven to be reliable.

13. Of course, Psillos holds that the term 'ether' referred to the electromagnetic field, but we've already seen why this won't help: Maxwell was as explicit as he could be that it was the ether *conceived as a material and mechanical medium in which electromagnetic waves were propagated* for which the existing evidence provided definitive confirmation.

14. Robison also reports that Black gave him "Newton's Optics to read, advising me to make that book the model of all my studies" (1803 vii). Interestingly, some scholars have suggested that it was in fact what Robison described as his "last Office of friendship"—his editorship of Black's work, which had gone almost entirely unpublished in Black's own lifetime—that is responsible for our comprehensive picture of Black as a "meticulous inductive philosopher" (e.g., Christie 1982 47ff).

15. Lavoisier displays the same rhetorical opportunism and keen sense of his own argumentative position in his antiphlogistic writings, where he acknowledges the hypothetical character of the material fluid theory of heat, but dispenses with any need for its defense simply by noting that this commitment is shared by his phlogistic opponents (McKie 1935 231, Morris 1972 31).

16. Indeed, Morris argues that regarding changes of state themselves as chemical transformations was "the most striking innovation in Lavoisier's theory of heat" (1972 34).

17. Here, of course, Lavoisier himself disputes Psillos's historical claim. I have, however, chosen to focus our attention on the particular scientists Psillos cites as exemplifying the cautious judgment that the caloric theory was not sufficiently supported by the evidence to be believed, rather than on the important endorsements and confirmational judgments in favor of the material fluid theory of heat by earlier and contemporary luminaries, such as Boerhaave, 'sGravesande, Franklin, Brisson, Homberg, and Monge, among others.

18. In a useful general discussion of the reception of Mendelism in England, Olby also points out how natural it was for later thinkers like William Bateson to treat even the inheritance of *discontinuous* characters as a special case of Galton's Ancestral Law involving prepotency (1987 413).

19. Cf. Galton (1987 402): "The same statement may be put into a different form . . . by saying that each parent contributes on average one-quarter, or $(0.5)^2$, each grandparent one-sixteenth, or $(0.5)^4$, and so on, and that generally the occupier of each ancestral place in the *n*th degree, whatever be the value of n, contributes $(0.5)^{2n}$ of the heritage."

20. Again, if we strain to find some correspondence we might note that the fraction associated with each *generation as a whole* in the Ancestral Law turns out to correspond to what contemporary theory regards as the coefficient of relatedness for each *individual member* of that generation to the organism in question, but the equation specified by the law itself is still a sum, and contemporary genetics simply does not recognize an organism as inheriting anything from its ancestors that is summed together in the way specified by this equation.

21. In *Jacobellis v. Ohio* (1964, concurring opinion).

8

Science without Realism?

> We should be careful to get out of an experience only the wisdom that is in it—and stop there; lest we be like the cat that sits down on a hot stove lid. She will never sit down on a hot stove lid again—and that is well; but also she will never sit down on a cold one any more.
>
> —Samuel Langhorne Clemens (Mark Twain), *Pudd'nhead Wilson's New Calendar*

The first five chapters of this book suggested not only that the problem of unconceived alternatives constitutes the most serious *prima facie* challenge for scientific realism but also that we have strong historical evidence of our repeated vulnerability to this challenge. That is, we have repeatedly failed even to conceive of equally well-confirmed alternatives to our best theories that were sufficiently scientifically serious as to be actually accepted by later scientists and scientific communities. The last two chapters argued that realist responses to challenges grounded in the historical record offer little reason to resist even the original pessimistic induction, much less the problem of unconceived alternatives itself. That is, we have found no prospectively applicable way to distinguish particular present theories (or parts thereof) from their rejected predecessors in a way that should lead us to expect the ultimate fate of the former to differ from that of the latter, nor any reason to think it less likely that there are serious, well-confirmed alternatives to our own best scientific theories that remain unconceived by contemporary theorists. Moreover, when we simply consider the specific evidence we have in support of some particular theory, both our brute intuitions and our considered reflective judgments that the available evidence renders the possibility of such alternatives remote in some particular case are demonstrably unreliable. Perhaps, then, it is time to ask whether it is possible to live without scientific realism and what the scientific enterprise might look like if we tried. That is, we might well wonder if any coherent positive image of scientific inquiry and its products remains open to us if we abandon what we described in chapter 1 as educated common sense about science: the view that our best scientific theories simply tell us (probably and approximately) how things really stand in otherwise inaccessible domains of nature.

We need not approach such a question empty-handed. A long-standing minority tradition among scientists and philosophers of science alike has sought not only to explore the various challenges for scientific realism, but also to develop some alternative positive conception of the scientific enterprise itself. Most commonly, proposals for such alternatives have pursued some version of the idea that even our best scientific theories are simply tools or instruments for making empirical predictions and achieving other practical ends, rather than literal and/or accurate descriptions of otherwise inaccessible domains of nature, and such sentiments concerning scientific theories have a remarkably long intellectual pedigree. Indeed, Karl Popper's (1963 chap. 3) famous critique of this "instrumentalist" position as intellectually sterile counts Andreas Osiander (author of the unsigned Preface to Copernicus's *On the Revolutions of the Celestial Spheres*), Cardinal Bellarmino, and Bishop Berkeley as notable defenders of the view, even while resisting Duhem's claim to find its historical antecedents in classical Greek thinkers (for criticism of Popper's somewhat simplistic historiography, however, see Fine 2001). But even those who find themselves attracted to such an instrumentalist conception of science have remained deeply divided concerning just what status we are to ascribe to the claims of our best theories and how best to understand the thesis that they are no more than tools or instruments for accomplishing our practical objectives.

The nineteenth-century physicist Ernst Mach, for example, grounded his distinctive and influential version of this instrumentalist position in a radical phenomenalism. He insisted that "[w]hat we represent to ourselves behind the appearances exists *only* in our understanding, and has for us only the value of a *memoria technica* or formula" (1911 49); thus the sole object of science is its "economical office" of "replac[ing], or *sav[ing]*, experiences, by the reproduction and anticipation of facts in thought" (1893 577) and with "the *least possible expenditure of thought*" (1893 586). Mach argued, for instance, that "[i]n nature, there is no *law* of refraction, only different cases of refraction," and therefore Snel's "law of refraction is a concise, compendious rule, devised by us for the mental reconstruction of" innumerable individual experiences (1893 582). Similarly, he regarded theoretical hypotheses simply as devices for the systematic classification, summary, organization and coordinated expression and prediction of innumerable particular appearances. Although Mach recognizes the existence of imperceptible *degrees* or instances of observed phenomena, he argues that in atomic theory, "atoms are invested with properties that absolutely contradict the attributes hitherto observed in bodies," and thus that the atomic hypothesis is only "a mathematical *model* for facilitating the mental reproduction of facts"

(1893 589). Theoretical concepts like 'atoms', Mach insists, are merely "provisional helps," ultimately to be discarded in favor of more direct descriptions of phenomenological appearances—not because they seek unsuccessfully to describe a reality beyond appearances but rather because they *successfully* but only *indirectly* describe collections of coordinated and systematized experiences themselves.

By contrast, thinkers like Poincaré and Duhem were moved to instrumentalist sympathies by the same sorts of developments in physics around the turn of the twentieth century that provoked their concerns about underdetermination and the pessimistic induction (for more details, see Worrall 1982). By this time, the progress of physical science had begun to suggest that there might be quite genuine cases of differences between actual competing scientific theories that could not possibly be adjudicated by any straightforward appeal to empirical tests or observations. To use a famous example of Poincaré's (though not a case of actual competing theories), any set of measurements of the angles in a triangle marked out by appropriately oriented perfectly rigid rods can be accommodated by the assignment of any number of different combinations of underlying spatial geometries and compensating 'congruence relations' for the rods in question; if the sum of the angles differs from 180 degrees, for instance, we could either interpret the underlying geometry as Euclidean and conclude that the distance marked out by each rod varies with its position and/or orientation, or assume that the distance marked out by each rod remains constant and conclude that the underlying geometry of the relevant space is non-Euclidean. Poincaré's response to this threatened form of underdetermination was to embrace conventionalism; that is, he regarded such theoretical matters as the assignment of a particular physical geometry to space as matters of choice or convention to be decided on grounds of greatest convenience. And this in turn implied, he suggested, that the quite useful ascription of a particular geometry to space by a theory should not be construed as literally attributing *anything* (truly or falsely) to nature itself: "[T]he question: Is Euclidean geometry true? . . . has no meaning. We might as well ask if the metric system is true, and if the old weights and measures are false . . . One geometry cannot be more true than another; it can only be more convenient" (Poincaré [1905] 1952 50).

Duhem sometimes drew a similar moral from his own distinctive worries about various forms of underdetermination, as well as from concerns about the pessimistic induction and the role played by idealization in physical theories: at various points he suggests that "hypotheses [are] not judgments about the nature of things, only premises intended to provide consequences conforming to experimental laws" ([1914] 1954 39), and that "propositions introduced by a theory . . . are neither *true* nor *false*; they are only *convenient* or *inconvenient*" ([1914] 1954 334). On the other hand, both Duhem and Poincaré insisted explicitly that the skeptical attitude they grounded in the repeated and radical discontinuities between the theoretical conceptions of nature accepted at different times across the history of scientific theorizing was of limited scope. It applied *only* to our efforts "to penetrate beyond the teachings of experiment or... surmise realities hidden under data observable by the senses" (Duhem [1914] 1954 274) when in fact these "merely nam[e]... the images we substituted for the real objects which Nature will hide for ever from our eyes" (Poincaré [1905] 1952 161). That is, both thinkers retained full confidence in the "experimental laws" or generalizations about observable phenomena uncovered by scientific investigations, even as each denied that such investigations were able to disclose (or that "mathematical theories" described) the actual underlying constitution of nature itself. Poincaré concludes that "we must not ask from [mathematical theories] what [they] cannot give us" and that it would be "an unreasonable demand" to expect our theories "to reveal to us the real nature of things"; instead "[t]heir only object is to co-ordinate the physical laws with which experiment makes us acquainted..." ([1905] 1952 211). Duhem argues likewise that the only legitimate aim of science is to "represent as simply, as completely, and as exactly as possible a set of experimental laws" ([1914] 1954 19), and famously goes so far as to insist that our attempts to *explain* the empirical regularities we uncover are regarded as the proper purview of speculative metaphysics rather than empirical science. It is only when we try to "strip reality from the appearances covering it like a veil, in order to see the bare reality itself" ([1914] 1954 7), he insists, that neither our own nor any genuinely scientific theories can do what we ask of them.

Even this whirlwind tour through some instrumentalist themes in the writings of Mach, Duhem, and Poincare suffices to illustrate that it has been no easy matter to parlay the intuitive idea that scientific theories are simply tools or instruments for achieving practical aims into a single clear and compelling vision of the scientific enterprise. It is quite striking, however, that even some of these early efforts to articulate an instrumentalist account of scientific theories share an explicitly semantic or linguistic character: faced with scientific claims that seem to concern events, entities, or phenomena beyond direct experience or observation, instrumentalists have repeatedly insisted either that such claims have a meaning very different from the one they seem to have or, if not, that claims of this sort can be eliminated from science altogether. On what we might call Mach's "reductive" variety of instrumentalism, for instance, the very *meaning* of theoretical discourse is exhausted by the implications it has concerning our sense experiences, and we may therefore simply believe the central claims of our best scientific theories after all, once they are correctly interpreted as only "really" saying something about collections of phenomenological experiences or appearances in the first place. Duhem and Poincaré seem instead to regard many scientific theories as aspiring or attempting to describe an underlying, inaccessible reality and/or explain observable events or experiences by appeal to it and simply reject these ambitions as unsatisfiable or even unscientific in some way. Each proposes at some point a form of what we might call "syntactic" instrumentalism, on which theoretical claims are neither true nor false, nor even "really about" the world itself, but are instead simply convenient syntactic devices guiding our inferences from some observations or observable states to others. But each also at least suggests a quite different "eliminative" form of instrumentalism, on which such theoretical claims really do make just the descriptive claims about inaccessible domains of nature that they seem to but can nonetheless be eliminated altogether from our scientific descriptions of the world. Thus, instrumentalist sympathies have produced a wide variety of importantly divergent attitudes (sometimes within the works of a single author) toward the cognitive, semantic, and epistemic status of theories, including the view that extant scientific theories do not in fact make claims about inaccessible realities behind observable phenomena, that the scientific enterprise need not do so, and that it should not.[1]

Perhaps it is less surprising that such a generally linguistic or semantic strategy of analysis appealed to logical positivist and logical empiricist thinkers through the middle decades of the twentieth century. The sort of reductive instrumentalism favored by Mach would engage the sympathies of those positivists and early empiricists who sought to carry out a reduction of all scientific language to a privileged phenomenological or observational basis, a project pursued most notably by the early Carnap, but also influentially by Bridgman. And when this reductive enterprise encountered the apparently insurmountable technical obstacles that led it to be abandoned even by its original architects, some positivist and empiricist thinkers would retreat to syntactic instrumentalism's distinctive claim that theoretical propositions are devoid of any meaning at all beyond the license they provide to infer from one observable state to another. That is, in the spirit of some of Duhem and Poincaré's remarks, these thinkers argued that despite appearances to the contrary the claims of theoretical science are nonassertoric: they are not claims about what the world is like, they carry no straightforward ontological commitments regarding unobservable entities, and they do not possess truth values at all.

Of course, this syntactic form of instrumentalism demands an extremely counterintuitive view of the semantics of theoretical discourse. Our scientific theories certainly *seem* to be making actual descriptive claims about what the world is like and to be put forward with the intention of so doing: if they make the claim that material objects like tables and toasters are made up of atoms constituted by electrons orbiting nuclei, this seems for all the world to say something about the actual constitution of tables and toasters and not simply that some observable states of affairs may be safely inferred from others. Thus, many positivists would evade syntactic instrumentalism's implausible account of the semantics of theoretical claims by embracing the eliminative version of instrumentalism instead: equally in the spirit of suggestions we have noted from Duhem and Poincaré, these thinkers insisted that while theoretical discourse is both meaningful and irreducible, it can nonetheless be eliminated from science altogether. This eliminative form of instrumentalism gained further currency among logical positivists and logical empiricists following such technical developments as the invention of the Ramsey sentence and the proof of Craig's Theorem. The import of the latter seemed to be, as Hempel explained in introducing his famous "Theoretician's Dilemma," that "any chain of laws and interpretive statements establishing [definite connections among observable phenomena] should then be replaceable by a law which directly links observational antecedents to observational consequents" ([1958] 1965 186). But the significance of Craig's theorem was immediately controversial: perhaps most influentially, Earnest Nagel (1961 136–137) pointed out several features of the theorem ensuring that it would be of quite limited relevance to the eliminability of actual theoretical discourse from science. And more recently, the profound differences between actual scientific theories and the sorts of artificial formal systems to which tools such as Craig's Theorem and Ramsey's technique apply have led these results to be regarded as having dubious relevance for the genuine prospects of instrumentalism.[2]

Thus, recent history has not offered much encouragement to those who would escape realist commitments either by reinterpreting the central claims of our scientific theories or trying to remake those theories so as to eliminate such claims. Philosophers of science have encountered what seem to be insuperable obstacles at every turn in trying to show either how the claims of theoretical science can be reduced or interpreted away or how science can be sanitized so as to remove claims about theoretical and/or unobservable entities from it. The challenges facing any such linguistic or semantic strategy for developing instrumentalism now seem sufficiently serious and its compensating attractions sufficiently slight that most contemporary philosophers, not to mention scientists themselves, regard such an enterprise as somewhere between hopeless and bankrupt.

Despite the historical influence they have enjoyed, however, such linguistic or semantic approaches to instrumentalism simply do not exhaust the available options for trying to develop the idea that even our best scientific theories are tools or instruments for guiding our practical engagement with the world rather than literal and/or accurate descriptions of otherwise inaccessible parts of that world. And we will here explore an alternative approach that is of quite a different character, although it is equally in the spirit of suggestive remarks offered by

Duhem and Poincaré as well as others since. On this alternative, we might allow that our scientific theories really do make just the descriptive claims about inaccessible reaches of nature that they seem to make, and even that the fruitful investigation of nature by human beings typically *requires* that we propose and explore claims of this character, but nonetheless insist that we can make perfectly good practical use of the claims of such theories without believing what they say about the natural world. This approach might be called "epistemic" by contrast with earlier "semantic" or "linguistic" approaches to instrumentalism, because it restricts the set of *beliefs* to which we regard ourselves as entitled by the dramatic empirical successes of our best scientific theories, rather than restricting or reinterpreting the very claims that those theories are understood to be making in the first place. And the fundamental commitment of this distinctively epistemic form of instrumentalism would seem to be shared (although without this suggestive unifying label), by any number of more recently influential alternatives to scientific realism, including those developed by Thomas Kuhn, Bas Van Fraassen, and Larry Laudan.

Arthur Fine offers a useful general characterization of this epistemic variety of instrumentalism in describing how its commitment to scientific claims falls short of endorsing their truth:

> Instrumentalism satisfices; it goes for something less than truth. In general, instrumentalism would be satisfied with the instrumental reliability (in the broad sense) of a theoretical story, and it treats reference in this same reliabilist way. Thus, we might say that instrumentalism treats scientific stories as though they were true, just in so far as it relies on them, and their postulated entities, as useful guides for whatever practical and theoretical jobs may arise.... One sometimes hears this expressed by saying that instrumentalism treats theoretical entities as fictions. But, if genes (for example) are to be called fictions, then one must be clear that they are not *mere* fictions, nor is genetic theory to be assimilated to some amusing piece of scientific fantasy. (1986 157)

The fundamental idea at work here would seem to be that we are to make use of our best scientific theories for practical purposes, but not to *believe* what those same theories tell us about the natural world. Although this suggestion might seem intuitively clear enough, closer scrutiny reveals that it will not suffice as an adequate general characterization of the epistemic instrumentalist approach. The central problem is that to use a theory for prediction, intervention, and other practical ends just *is* to believe at least *some* of the things it tells us about the world—for example, that the shuttle really will fall back to Earth if the booster rockets fail to ignite, or that a certain drug really will alleviate the symptoms of a given disease, or that we really should expect to find fossils of a particular kind in a particular geographic location. So the simple distinction between making use of a theory and believing what it says cannot deliver what the instrumentalist needs.

If the instrumentalist must believe at least some of what a theory says about the world in order to make effective instrumental use of it, it might seem natural enough to try to find some principled way to distinguish those claims she must believe in order to make use of a theory from those that she need not. And it would seem that our commitment to the instrumental reliability of a theory means (at least) that we trust the guidance it offers to our practical or pragmatic engagement with the world around us. That is, perhaps what it means to be an instrumentalist about any particular theory is to believe the *empirical predictions* and *recipes for intervention* that the theory offers, but not the *description* of some part of nature in which those pragmatic recommendations are grounded. This suggestion seems to reflect at least the spirit of some of the most influential recent proposed alternatives to scientific realism. Kuhn, for example, famously claims that a later scientific theory is typically "a better instrument for discovering and solving puzzles" than its predecessors, offering more impressive "puzzle-solutions and.... concrete predictions," but denies that this is because such later theories are "somehow a better representation of what nature is really like" ([1962] 1996 206). Likewise, Laudan denies that we should view successive theories as achieving greater truthlikeness or more closely approximating the truth about nature, but insists that the scientific enterprise is nonetheless progressive because successive theories typically improve over time in their ability to solve the various empirical and conceptual problems that we face (1977, 1996). To just the extent that Kuhn's "puzzles" and Laudan's "problems" involve practical contexts of prediction and intervention, these sentiments would seem to incorporate the approach to epistemic instrumentalism sketched above.

While this proposal shares the intuitive appeal of the idea that we might use our theories without believing them, it is ultimately no better able to provide a satisfying general characterization of the epistemic instrumentalist's position. For one thing, many of the predictions a theory makes just *are* descriptive claims about remote and inaccessible aspects of nature, and are thus *part of* its theoretical description of the world. Howard Stein poses this problem in a particularly elegant way, paraphrasing a remark of Eugene Wigner's pointing out that one also "uses quantum theory, for example, to calculate the density of aluminum" (1989 49). To take another example, among the most important predictions of our present cosmological theory are the characteristics it specifies for the field of so-called "dark energy" posited to explain the present expansion of the universe. Still more important, however, is the fact that a theory's predictions of even the most mundane and familiar events (like earthquakes, supernovae, or tumors) will be shot through with just the same theoretical apparatus it uses to describe the world quite generally: *what* the theory actually predicts is the shifting of the Earth's tectonic plates, the violent explosion of a distant star, or the rapid production and accumulation of genetically abnormal cells. That is, a theory's predictions of events, entities, or phenomena and its recipes for intervention with respect to them are typically offered in precisely those terms the theory uses to describe the world generally, and so cannot be more epistemically secure than the theory's own description of nature: we can hardly say that we believe a theory's prediction that a "supernova" will occur at a given time and place, for instance, but not what it says supernovae are. And it is far too late in the day to suggest that we might try to retreat into some kind of pure observation language that involves no theoretical commitments at all, using the theory to predict and intervene with respect to needle-readings, sensations, or colored patches in the visual field rather than earthquakes, supernovae, or tumors. There is no such pure observation language, and even if there were we could not use it to describe the sorts of perfectly familiar phenomena, entities, and events with respect to which even the instrumentalist thinks our theories really do allow us to predict and intervene successfully.

We might usefully return here, however, to Kitcher's point about the diverse ways in which the reference of various tokens of a natural kind term can be fixed on different occasions of use. Recall that Kitcher hoped to use this distinction to try to show how some tokens of a natural kind term could be referential while other tokens of the very same term were not. Some of Priestly's tokens of "dephlogisticated air," for example, had their reference fixed by his intention to refer to air once the substance emitted in combustion had been removed from it: since there is no substance emitted in combustion in the way he imagined, these tokens of the term simply failed to refer. On other occasions, however, Priestly's dominant linguistic intentions were quite different: Priestly used the term 'dephlogisticated air' to refer to oxygen, Kitcher suggests, when he conceived of it instead as the substance he had isolated whose inhalation was rendering his breathing particularly light and easy or the substance he 'exploded together' with 'inflammable air' to produce either water or nitric acid. Similarly, some of Fresnel's tokens of the term 'light wave' could refer to electromagnetic waves of high frequency, Kitcher suggests, while those whose reference was fixed by Fresnel's intention to talk about oscillations of the molecules of the ether simply did not refer at all.

In chapter 6 we noted that Kitcher's distinction offers cold comfort to the scientific realist because it runs afoul of the "trust" argument. That is, it allows Priestly to refer successfully using the term 'dephlogisticated air' while having a radically mistaken theoretical description of this substance, which simply divorces the question of successful reference from the real issue concerning realism: whether or not we should trust the fundamental descriptions of nature offered by our best scientific theories. But Kitcher is nonetheless quite right to point out that the reference of many tokens of 'dephlogisticated air' simply didn't depend on the descriptive accuracy of Priestly's phlogiston theory in any significant way. When Priestly noted that the dephlogisticated air he had isolated rendered his breathing particularly light and easy, nothing in this claim depended on the accuracy of his chemical theory for the successful reference of its central terms and the claim is itself perfectly intelligible in a way that is independent of the theoretical descriptions Priestly would have associated with them. What Kitcher's distinction points out, then, is that our theoretical descriptions of the natural world, although ubiquitous in their influence, are not always or uniformly involved in the same way when we characterize events, entities, and phenomena in the world around us and the relationships between them.

These considerations suggest a natural way to characterize at least somewhat more precisely the suggestion that we might use our theories for prediction, intervention, and other pragmatic purposes without believing the theoretical descriptions they offer of the natural world. Though we cannot make use of our scientific theories for the pursuit of our practical endeavors without believing some part of what they say, perhaps it is open to us to believe the claims about entities, events, and phenomena that they make *as those very claims can be understood independently of the theory or theories toward which we are adopting an instrumentalist stance*. Take, for example, the claim that "the bithorax phenotype in *Drosophila melanogaster* is caused by a single mutation in the HOM complex of homeobox genes." This highly theory-laden claim carries a specific set of implications concerning entities, events, and phenomena we would expect to find in particular contexts *as we understand them independently of contemporary genetic theory*: that when particular materials from the bodies of organisms exhibiting a distinctive anatomical characteristic are subjected to any one of a specified set of laboratory procedures they will also exhibit a distinctive pattern on the resulting autoradiograph or electrophoresis gel, that a particular complex pattern of statistical dispositions will govern the appearance of that trait in the organism's descendants, and so on. This is certainly not to make the operationalist suggestion that this is what it "really means" to say that bithorax is caused by a single mutation in the HOM complex. Nor is it to suggest that we could ever hope to give any account either of what the various laboratory procedures for identifying the presence of a particular gene have in common or of the open-ended list of possible exceptions to usual patterns of Mendelian inheritance without making use of the theory itself. To construe theoretical claims only in terms of their theory-independent practical consequences in particular concrete contexts is simply not an effective way to try to think about nature; it is instead a terrible way to do so, and this is part of why the instrumentalist insists no less than the realist that we should be using our best scientific theories for this task instead! Identifying the various ways in which a theory's concrete consequences can be grasped independently of the theory itself is important only when we seek to specify the beliefs about the world to which we remain committed once we recognize that we are not epistemically entitled to belief in a particular theory even though we know of no better way to pragmatically engage a given aspect or domain of nature than by means of it.

Of course, if a theory makes claims about entities or events to which we have no access independently of the theories that describe them or claims for which there is no sense to be made in a way that is independent of those descriptions, the instrumentalist will not believe them at all, but will simply make use of them in the process of generating further predictions and recipes for intervention whose content does not depend exhaustively on the theory's own descriptive apparatus in this way. Contemporary particle physics, for example, does not permit quarks to be isolated and so posits gluons to bind quarks within a proton, but we have no point of contact with gluons that is independent of the theory that posits them in the way that Priestly did with his dephlogisticated air or contemporary molecular geneticists do with the *Drosophila* HOM cluster. Thus, the instrumentalist will not believe *any* of the claims of contemporary particle physics regarding gluons, but will nonetheless be perfectly happy to make use of such claims in the course of generating further predictions and recipes for intervention concerning phenomena with which she does have some contact or commerce in a way that is independent of the theory itself.

This conception of instrumentalism seems to me quite closely connected to a point Fine makes in emphasizing why the instrumentalist need not accord any special epistemic significance to the distinction between observables and unobservables in articulating her position. As he notes, the distinctive fundamental commitment of instrumentalism is simply to the *reliability* of a given causal story, whether that story concerns observable or unobservable entities:

> For according to instrumentalism what we want from our theories, posits, ideas, etc. in all the various contexts of inquiry is instrumental reliability. That is, we want them to be useful in getting things to work for the practical and theoretical purposes for which we might put them to use. This is the guiding pragmatic ideal of instrumentalism, and it treats all entities (observable or not) perfectly on par.... Of course if the cause happens to be observable, then the reliability of the story leads me to expect to observe it (other things being equal). If I make the observation, I then have independent grounds for thinking the cause to be real. If I do not make the observation or if the cause is not observable, then my commitment is just to the reliability of the causal story, and not to the reality of the cause. (1991 86)

The attraction of this picture is that it offers the instrumentalist a single systematic and principled commitment to the mere reliability of *all* of the claims made by a given scientific theory: it allows differences in our further beliefs to emerge simply from the fact that such reliability will be manifested differently (and we will therefore be able to accumulate different sorts of further evidence) for claims about observable rather than unobservable entities. On Fine's account, if the distinction between observables and unobservables is itself vague or indeterminate or a matter of degree, this might well impact the further beliefs the instrumentalist (or any of us) is able to form on the evidence of our senses, but it will not even be relevant to the fundamental commitment *of her instrumentalism*, which is simply to the reliability of the scientific theory in question across the board. In a like manner, we may resist the natural but ultimately misguided temptation we noted earlier to divide up the theory's own claims into those we need and need not believe in order to make effective practical use of it. What we want to say instead is that our fundamental commitment to the reliability of a given theory *in turn* commits us to the truth of whatever implications it may have for entities, events, and phenomena as they are conceived of outside of the theory itself (and indeed outside of all those theories toward which we are adopting an instrumentalist attitude). But of course, this is simply a generalization of Fine's approach to observability: when our theories carry implications concerning features or aspects of entities, events, and phenomena to which we have some independent route of epistemic access, our belief in the instrumental reliability of those theories leads us to expect these implications to turn out to be strictly and literally true. We will trust the remainder of a theory's implications to be reliable as well, but this will not mean believing any specific empirical claim about the world itself; it will mean only that they can be effectively used in generating further implications concerning matters (like the lightness of our breathing or the pattern on an electrophoresis gel) that can be ascertained in a way that is independent of the theory in question. The difference is not one of the epistemic attitudes we take toward the various claims of the theory itself, but rather of what *other* routes of epistemic access we may have to the furniture of the natural world about which the theory makes its presumptively reliable claims.[3]

This is emphatically *not* to say that the instrumentalist must or even could characterize the claims about successful prediction and intervention that her best theories lead her to believe in a way that is independent of *any* theory whatsoever, much less in the terms of a mythical "observation language." Suppose, as Quine suggested long ago, that the middle-sized objects of our everyday experience are no less "theoretical" entities hypothesized to make sense of the ongoing stream of experience than are atoms and genes, notwithstanding their greater familiarity. Then, it is by means of theories in this broadest sense that we come to have any picture at all of what our world is like, what entities make it up, and how they are related to one another. But recall that only some of these theories will be open to the distinctive challenge posed by the problem of unconceived alternatives that I have suggested lies at the heart of any serious objection to scientific realism. As we have noted from the outset, what Quine calls the hypothesis of the bodies of common sense does not appear to be vulnerable to this challenge and its descriptive apparatus is therefore available to the instrumentalist to use in characterizing the earthquakes, supernovae, and tumors with respect to which she believes her best scientific theories allow her to predict and intervene successfully. It is ultimately because Priestly's claim that inhaling the "dephlogisticated air" he isolated rendered his breathing light and easy for several hours can be quite naturally interpreted in terms of these bodies of common sense that we can and do continue to count it true long after we have rejected his phlogiston theory of chemistry as profoundly mistaken: notice that if a radical revision of *contemporary* chemistry were to deliver the verdict that that the substance Priestly inhaled was not what we think of as oxygen either (and indeed that there is no such thing), we would *still* have no trouble understanding Priestly's claim. Furthermore, it seems that this would be true even if we ultimately came to see this description of the events as radically mistaken, for instance by rejecting the idea that the "dephlogisticated air" Priestly isolated was even a substance. Likewise, if a radically different successor theory ultimately replaces contemporary genetics, we might find ourselves unable to endorse or even know how to evaluate the perfectly general claim that bithorax in *D. melanogaster* is caused by a mutation in the HOM complex of homeobox genes, but we will have no trouble understanding or endorsing our earlier claims about the banding patterns displayed on autoradiographs and gels and about statistical dispositions governing the reappearance of bithorax in the offspring of organisms exhibiting the trait even if we do not continue to speak in these terms.

Although this account of the matter rejects as unhelpful the very idea of a pure observation language or a foundational epistemic role for observability as such, it does help to explain why observation and other perceptual processes will typically play an important role for instrumentalists in fixing beliefs about the natural world. This is because our various sensory modalities will characteristically be among the routes most commonly used by us to secure a grasp of entities, events, and phenomena in ways that are independent of the theories toward which we adopt instrumentalist attitudes. Indeed, this role for our perceptual abilities is itself a part of the everyday hypothesis of the bodies of common sense as we invoke it to account for the experiences we have, and this is surely at least part of why scientists sometimes work quite hard to devise ways to render the entities posited by their successful theories accessible to more familiar processes of observation or detection. But there is no bright line to be found here between "observable" and "unobservable" entities or "observational" and "theoretical" terms and no foundational epistemic distinction between observational and other modes of access to the phenomena. Indeed, scientific investigation itself has revealed a wide variety of ways in which our own perceptual faculties can be misled or deceived and circumstances in which our ordinary experience of the world seems to tell us things that are simply not to be trusted. These failings of our perceptual abilities can be demonstrated in ways that do not depend on the theories that we use to discover them, just as a microscope's ability to increase optical resolution can be demonstrated independently of the theories we use to construct it: human beings can be shown experimentally to be poor judges of distance, time, and speed, for example,[4] and we have learned to trust our demonstrably reliable instruments over our demonstrably unreliable phenomenological experience to judge these matters *in the world given to us by our (theoretical) description of the bodies of common sense itself.* Thus, the only epistemic privilege here attached to sense experience is the broadly empiricist insistence that the evidence of our senses is where we must begin in finding out what the world is like.

It is also worth noting that the development of any such "hypothesis of the bodies of common sense" is itself an ongoing enterprise. It is by no means equivalent to a "folk physics" or any other folk science, if such there be. Such "folk" theories are manifestly false accounts of the workings of some part of the natural world and they stand refuted by the empirical evidence available in the domain of the theory long before we concern ourselves with the problem of unconceived alternatives—but of course we think that the hypothesis of the bodies of common sense is true! Particular people can and surely do believe false things, but the hypothesis of the bodies of common sense neither deifies commonplace collective wisdom about the external world nor simply replaces it with the descriptions found in our best scientific theories: it is rather common sense refined and sophisticated to reflect *everything* we know about the world of everyday objects as we experience and interact with them. This hypothesis begins with the fundamental supposition that our sense experience is produced by an external world full of enduring objects that we perceive and with which we causally interact, and it is this supposition that seems to escape any serious historical challenge from the possibility of radically distinct unconceived alternatives. But this commonsense grasp of the familiar objects in the world around us is continuously revised in response to discoveries about it prompted by developments in the sciences themselves: that objects dropped from moving vehicles will in fact fall along parabolic trajectories rather than directly toward the earth, that whales share important anatomical similarities with other mammals and not with fish, that the people around us constantly reason in ways that do not respect basic principles of rational inference.[5] Many such discoveries about the objects and interactions in the world of our everyday experience would almost certainly not have been possible but for the pursuit and development of self-consciously theoretical natural science, but we do not depend upon the sciences in order to understand the facts they report about the world of our everyday experience. Thus it is perfectly open to the instrumentalist to allow that our scientific theories are extremely powerful tools for successfully guiding our predictions and interventions with respect to earthquakes, supernovae, tumors, and the like as those entities are grasped through our everyday experience of the world, even while recognizing that our systematic knowledge of that everyday experience is itself refined, improved, sophisticated, and corrected by the further scientific investigations we have undertaken.

If we allow that our grasp of the world is, in this sense, "theories all the way down," it might well be impossible to coherently adopt instrumentalism as an epistemic attitude toward any and all theories whatsoever. That is, if we grant that all of our efforts to understand the world around us and our place in it are fundamentally theoretical in the broadest sense of the term, then it may well be that there is no coherent way to adopt an epistemic instrumentalist attitude toward all of our theories simultaneously.[6] It is fortunate indeed, then, that our epistemic position in no way demands this. But it is also far from clear that a suitably sophisticated version of the hypothesis of the bodies of common sense is the *only* theoretical resource that will remain available to the thoughtful instrumentalist in trying to characterize what she strictly and literally believes about the external world. The preceding chapters have been concerned to argue that the problem of unconceived alternatives is a genuine challenge for scientific realism and to expose both the breadth and depth of that challenge, but they have certainly not established that the challenge can never be overcome or that every single scientific theory must automatically fall victim to it. Though we have yet to encounter them, perhaps there are specific reasons to believe that some particular theories do not remain vulnerable to the same problem of unconceived alternatives that has plagued so many of their predecessors and contemporaries. If so, these theories will count among the descriptive resources to which a thoughtful instrumentalist can appeal in characterizing the claims about the world that she believes to be true literally and without qualification. If not, the instrumentalist will have to frame what she actually believes (when called upon to do so) in terms of the entities, events, and phenomena familiar from our everyday experience of the middle-sized bodies of common sense. Indeed, it is in some sense only now that the really difficult work begins: the work of trying to decide whether there are cases or contexts of inquiry in which the available evidence is sufficient to rebut the historically established presumption in favor of the existence of serious unconceived alternatives to even our most successful scientific theories.[7] This is to some extent an explicit invitation to establish realism piecemeal, by showing that our most general reason for rejecting scientific realism about theories can be defeated in particular cases. Although the two preceding chapters have tried to show why the sorts of evidence to which realists have traditionally appealed to distinguish the expected fates of present theories from those of their failed predecessors are simply not up to this job, the naturalistic spirit of our inquiry requires that we remain open to the possibility that we will find special reasons to doubt that some particular scientific theories remain vulnerable to the problem of unconceived alternatives that seems to attend our scientific theorizing quite generally. I remain cautiously optimistic that there will be something general to be said about our systematic vulnerability to the problem of unconceived alternatives (and we had better hope that there is, since our intuitions in particular cases concerning whether we are vulnerable to the problem or not have turned out to be so spectacularly unreliable), but this strikes me as important, outstanding, and much-neglected work facing the contemporary philosophy of science.

What the instrumentalist must be able to do, then, is connect the descriptive and inferential machinery of the theory toward which she adopts an instrumentalist attitude to entities, events, and phenomena as they can be characterized independently in the terms of *whatever* theories she strictly and literally believes, so that the former can be used to guide our pragmatic engagement with the latter. To adopt an instrumentalist attitude toward the phlogiston theory of chemistry, for example, we must know how the phlogistonian chemist would characterize parts of the world that are otherwise familiar to us in terms of her preferred theoretical descriptions, how she would use the theory's descriptive and inferential machinery to move from its characterization of one concrete situation to another, and how the resulting theoretical characterization could again be connected to a description of some part of the world that we are able to strictly and literally believe. This is how the instrumentalist phlogistonian arrives at the belief that heating the reddish powder in the jar by the door will force air into a connected enclosed vessel filled with water and that this air will support subsequent respiration and combustion longer and/or better than ordinary atmospheric air: such claims can be embraced without believing any part of the phlogiston theory itself or accepting any of the distinctive theoretical characterizations it offers of the central notions involved. Thus conceived, a merely instrumentally useful theory is not a device for taking us from sensations to sensations or observables to observables, as instrumentalists have traditionally supposed, but instead from states of affairs characterized in terms we can strictly and literally believe to other such states of affairs. And it is in this precise sense that we can recognize that the intervening theoretical machinery is neither reducible nor eliminable without believing that the claims it makes about nature are true. If I can say how the belief in a theory would lead me to predict and intervene with respect to entities, events, and phenomena as they can be characterized independently of that theory (and in terms of one or more theories that I do strictly and literally believe), then I know what further beliefs I am committed to by thinking that the theory is instrumentally useful but not true.

If this view of the matter is right, the characteristic epistemic attitude at the heart of such instrumentalism is not only already familiar to us, but is also almost strictly identical to the attitude the realist herself adopts toward some particular scientific theories. Thus, we might ask what the scientific realist means when she allows that we can perfectly well make use of Newtonian mechanics to send rockets to the moon without believing the theoretical description of nature that it gives. She will, of course, completely reject the account of nature that Newton offered—gravity is not a force exerted by massive bodies on one another, there is no absolute space or time, and so on. But she nonetheless knows perfectly well how a Newtonian would apply the theory to make predictions about and intervene in the natural world *of entities, events, and phenomena as they are given to her by other theories that she does believe*. That is, the relativity theorist has an independent account of the cannonballs, inclined planes, and rockets to which she could apply the theoretical mechanics she believes to be literally true, but she also knows how a Newtonian would apply her own distinctive theory to characterize *those very* cannonballs, inclined planes, and rockets in terms of the masses, forces, collisions, etc. that would allow her to predict and intervene with respect to them. This has nothing to do with the fact that cannonballs, inclined planes, and rockets are observable entities: she knows equally well how a Newtonian would identify forces and masses so as to make predictions about the gravitational motions of subatomic particles. And over whatever domain she thinks the theory is (more or less) instrumentally reliable she can make use of it because she knows how to apply it like a Newtonian would to entities, events, and phenomena whose existence she countenances independently of the theory. But of course the instrumentalist, too, knows how the Newtonian (or the relativity theorist, or the molecular geneticist) will connect the terms of her favored theory with the entities, events, and phenomena that the instrumentalist herself independently recognizes and accepts: at least those familiar from her everyday experience of the world mediated by the hypothesis of the bodies of common sense, and perhaps more besides. Thus, it is precisely because the instrumentalist knows how a true believer would connect up the atoms, genes, and spacetime curvature of her scientific theories to the rockets, rabbits, and autoradiographs of her own everyday experience in innumerable concrete ways that she knows precisely what it is that she believes when she adopts an instrumentalist attitude toward those same theories. And she will agree that molecular genetics, or general relativity, or whatever, is an extremely reliable tool for guiding her engagement with those (independently-characterized) objects without believing that it is therefore true. But of course this is just what the realist does in connecting the distinctive descriptive and inferential apparatus of Newtonian mechanics to entities, events, and phenomena she thinks are truly described by some other theory or theories and using Newtonian mechanics to predict and intervene with respect to them. That is, the instrumentalist will take precisely the same attitude that the realist applies in particular contexts to theories she has specific reasons to disbelieve (like Newtonian mechanics) or to theories that she does not know how to interpret realistically (like quantum mechanics) and simply apply that familiar attitude much more broadly than the realist does.[8] Indeed, she takes the historical evidence for the problem of unconceived alternatives as a sufficient reason to apply this same familiar attitude to virtually every theory concerning the fundamental constitution of and dynamical principles at work in nature, absent a showing of some specific reason to refrain from doing so.

As this comparison suggests, it was perhaps always a mistake to think that our task was to identify some crucial difference between the epistemic attitudes of realists and instrumentalists toward theories or theoretical claims or theoretical knowledge as such. The characters traditionally identified as the realist and the instrumentalist both recognize theories that they strictly and literally believe to be true and theories that they think are merely instrumentally useful over a wider or narrower domain of nature. The instrumentalist simply assigns a much larger set of the theories we actually have to the latter category. Furthermore, she is willing to assign a particular theory to the latter category even when she has no competing account of the same natural domain that she assigns to the former instead: the instrumental utility of the theory requires simply that she be able to connect its descriptive and inferential machinery to some set of entities, events, and phenomena in which she strictly and literally believes as they can be characterized outside of that theory itself. The difference between the realist and instrumentalist, then, is not one of a global epistemic attitude toward "theories," but rather a local difference in the specific theories each is willing to believe on the strength of the total evidence available. This contrast also makes clear that there is no privileged or foundational role for the hypothesis of the bodies of common sense or any specific hypothesis *built into* the instrumentalist's account of the world and our knowledge of it. Instead, it simply turns out that this particular theoretical account can survive the challenge posed quite generally by the problem of unconceived alternatives, and it is therefore available to the instrumentalist (perhaps along with other particular theories as well) for characterizing what she strictly and literally believes, including the consequences or implications of those theories toward which she adopts a merely instrumentalist attitude instead.

While I have suggested that the characteristic instrumentalist attitude is familiar from the realist's own approach to theories like Newtonian mechanics, there will still be important differences between the epistemic position of a person who adopts such an instrumentalist attitude toward most or all of our leading scientific theories and that of one who does not. In most cases, the realist believes that the best-confirmed account we currently have of a given natural domain is (probably, approximately) true, so she reserves the instrumentalist attitude for theories that have useful ranges of practical application despite the fact that they conflict with a better-confirmed account and so are presumed to be false. When the realist uses her theories to try to discover new phenomena or make novel predictions in a given scientific domain, then, she will typically make use of a theory she thinks is true and not merely instrumentally useful for this purpose. By contrast, a person who adopts the instrumentalist attitude concerning most or all of our scientific theories must be willing to accept this attitude toward a particular theory even when she has no better-confirmed account of the same natural domain to offer in its place. She will thus find herself discovering new phenomena and making novel predictions about the natural world using a theory that she does not think is true. But it does not seem that any subtle incoherence lurks in applying the sort of instrumentalist attitude the realist herself embraces in some cases to a theory of a given natural domain when we have no competing account of that domain that we strictly and literally believe to be true. It seems perfectly coherent to deny that our own best scientific theories offer us even approximately true accounts of how things stand in otherwise inaccessible domains of nature, but insist that they are nonetheless the best *presently available* places to look for guidance in planning our own engagement with the natural world: in trying to produce drought-resistant crop plants, in trying to anticipate the distribution and characteristics of undiscovered fossils, in trying to send rockets to the moon.[9]

Furthermore, the epistemic instrumentalist will often be justified not only in exploring but also in *believing* the implications of our best existing scientific theory in such novel applications. After all, in typical cases such a theory has already proven to be a reliable guide to prediction and intervention across a wide range of phenomena in a given natural domain that are systematically related in intricate detail, and it is this track record that grounds whatever degree of confidence the instrumentalist has in the continued reliability of the theory with respect to novel applications and predictions of further phenomena in that same systematically related domain (for further discussion, see Stanford 2000). Although she suspects that the theory will *ultimately* be replaced by a still more instrumentally powerful alternative that is presently unconceived, she has no specific reason to think that any *particular* novel application of a theory will be one in which its demonstrated reliability with respect to the phenomena in its purview will founder.[10] And, of course, she knows of no other ground for her expectation about the outcome of some novel case with even nearly as much to recommend it as that offered by her most instrumentally powerful theory.

Furthermore, the instrumentalist knows that the most effective strategy available for trying to actually reach any presently unconceived instrumentally superior alternative will typically be articulating, exploring, extending, and testing the best theory she currently has. Indeed, we must not make the mistake of imagining that a theory is *simply* a tool for moving inferentially from some states of affairs we strictly and literally believe to others: this would be, as Stein insists, to make "a false estimate of the scope of a theory *as an instrument*" (1989 49). We have already noted that the instrumentalist recognizes our leading scientific theories as the best conceptual tools we have *for thinking about nature*, and this recognition encompasses the fact that they are the most powerful resources available to us in extending, pursuing, and expanding our further inquiry into the natural world itself:

> The instrumentalist himself, if he wishes to do justice to the role actually played by theories in science, had better extend still further his conception of what theories are instruments *for*, to include their role as resources for inquiry; especially as sources of clues in what Peirce called "abduction": the search for good hypotheses. (Stein 1989 52)

That is, the sophisticated instrumentalist recognizes that our leading scientific theories are the best foundation and starting point we have not only for uncovering new and unexpected phenomena, but also for opening up new areas and paths of inquiry, and in guiding ourselves to the even more powerful conceptions of natural domains that will ultimately replace the ones we now have.

We might illustrate this point in connection with the instrumentalist's more general convictions by returning to the case of Weismann's theory of the germ-plasm, for there is no question that systematically exploring, challenging, and defending this view turned out to be critical to developing many features of the subsequent theories of inheritance and ontogeny that would ultimately come to displace it. Of course it seems quite right to endorse what would have been the instrumentalist's judgment at the time: that Weismann was in possession of a powerful conceptual tool for conceiving of and guiding our practical engagement with a particular domain of natural phenomena. This tool directed our efforts at prediction and intervention effectively with respect to a wide variety of the phenomena of inheritance and generation, from patterns of individual variation and reversion to the formation of sex cells (including Weismann's famous early prediction of the need for reduction division) to the course of embryological development. In this respect it offered some genuine advantages over the existing alternatives; it alone, for example, seemed able to allow simultaneously for both nuclear control of the cell and cellular differentiation. These advantages and the theory's systematic practical successes surely justified Weismann's willingness to make use of the theory in pragmatic contexts of prediction and intervention with respect to further phenomena of inheritance and development: for example, to insist (rightly) that acquired characteristics cannot be inherited and to predict (mistakenly) the existence of a special accessory germ-plasm in cells capable of regeneration.[11] But for all that, an instrumentalist contemporary would have been equally right to think that Weismann was not in possession of even an approximately accurate description of the domain whose constitution and mechanics he sought to describe. The theory of the germ-plasm was systematically mistaken in any number of its most fundamental features, not the least of which were those respects in which we have noted Weismann failed to conceive of important later alternatives: that the germ-plasm is an expendable rather than productive resource for the cell which must be disintegrated over the course of ontogeny and used up in effecting cellular differentiation, that cells with different ontogenetic fates must contain different controlling germinal materials, that germ-plasm must be reserved in a special inactivated state not only for inclusion in the sex cells but also to sustain the permanent possibilities of regeneration and dimorphism, and so on. And one of the most important constructive roles for the theory was its suggestive implications regarding these further phenomena and further directions of research that would prove fruitful, not only for refining the theory and improving our practical abilities of prediction and intervention, but also for opening up the conceptual space of theoretical alternatives that would ultimately lead to the replacement of the theory itself.

Likewise, the epistemic instrumentalist of the present day will suggest that we should certainly allow that our own best theories are effective guides to our practical projects of prediction and intervention and that we are justified in using them rather than known inferior alternatives to structure our engagement with a wide variety of natural phenomena, but she will remain convinced that they can accomplish all of this (just as Weismann's germ-plasm theory did) while being profoundly misguided about fundamental aspects of the constitution and dynamics of the domains of nature they describe (just as Weismann's theory was). She will insist that our inability to exhaust the space of serious and well-confirmed theoretical possibilities has left us with many theories that are neither literally nor even approximately true, but that are nonetheless able to serve us extremely well in our practical engagement with the world even as they guide the further research that will enable us to expand our view of the theoretical possibilities and ultimately lead us to replace existing accounts of nature with even more powerful successors.

Indeed, there is a crucial further difference between instrumentalism and realism in that we should expect the instrumentalist to be far more willing to devote serious time and effort to the exploration of fundamentally distinct alternatives to our current theoretical conceptions of nature, as opposed to simply working to sophisticate or supplement our best current theories. For the realist, such exploration will, in general, have quite limited interest or motivation, for its rationale is found only in the relatively small chance she recognizes that contemporary theories might ultimately turn out to be fundamentally mistaken. By contrast, the instrumentalist fully expects even the best current theories to ultimately be replaced by even more powerful and fundamentally distinct conceptual tools for thinking about nature, and so she has every reason to invest serious time and effort in searching for these more powerful successors. In this sense, the instrumentalist is substantially more committed to the open-ended character of scientific inquiry than her realist counterpart.

When instrumentalism is conceived in this robust manner, there would seem to be little reason left to endorse Popper's claim that it is a:

> narrow and defensive creed according to which we cannot and need not learn or understand more about the world than we know already. A creed, moreover, which is incompatible with the appreciation of science as one of the greatest achievements of the human spirit. (1963 103)

The form of epistemic instrumentalism I have suggested does not propose that we simply follow along after our realist aspirations have given us powerful theories and timidly restrict our beliefs to their immediate practical utility in a lawyerly fashion. Its conception of inquiry is neither intellectually stultifying nor dismissive of the very real achievements that scientific theorizing has unquestionably delivered to us. Instead, the epistemic instrumentalism we have considered recognizes that boldly proposing hypotheses about the inaccessible workings of nature and then exploring and testing their implications has served as the preeminent engine of intellectual progress in the modern age. It acknowledges that scientific theorizing has produced and shows every promise of continuing to produce increasingly powerful and sophisticated tools for guiding our engagement with nature, and it regards us as eminently justified in continuing to explore the natural world by searching for more and better conceptual tools of just this same sort in just this same way. The epistemic instrumentalist will insist no less than the realist that we continue to challenge our best scientific theories by uncovering and testing further empirical implications they have, that we use them to unearth new phenomena and new ways to predict and intervene in the course of events around us, that these theories serve as the appropriate starting point in trying to determine how they themselves can be refined, improved, and developed, and perhaps even that we work to unify our various scientific theories with each other and with whatever else we believe to produce a single coherent, consistent, and systematic account of the natural world as a whole.[12] She will not demand that a theory meet some looser standard of evidence in order to be accepted, or pursue the further implications of our theories less doggedly, or invest those implications with less significance, than the realist, for she is after the most powerful, most stringently tested, most broadly applicable, and most predictively accurate set of conceptual tools that are to be had. In short, the instrumentalist is in a position to take the claims of our best scientific theories about nature every bit as *seriously* as the realist does, even as she declines to believe everything about the world that the realist believes.

At the end of the day, then, perhaps we should worry instead that taking the claims of our scientific theories so seriously threatens to turn epistemic instrumentalism into a position that is only verbally distinct from scientific realism itself, as Nagel (1961 139) once famously suggested concerning instrumentalism of the distinctively semantic or linguistic variety. Such a claim will appeal only to those who share the mistaken conviction that an embrace of the realist label is the only possible way to express their enthusiasm for and commitment to the importance of the scientific enterprise, rather than the belief that the present accounts of nature produced by that enterprise are probably and/or approximately true. For any realist who embraces the substantive features of the view developed here will find that she has given up the central commitments animating the realist account of scientific inquiry and its products. To be sure, the realist and instrumentalist agree in embracing the theoretical proposition that there is an external world full of rocks, spiders, and distant stars as well as in any number of further beliefs about the entities, events, and phenomena in that world whose truth do not depend on the accuracy of our more detailed scientific theories: that the rocks are hard, that the spiders spin webs, and that the distant stars shine at night. They also agree that our scientific theories are the very best tools we have for predicting and intervening with respect to these rocks, spiders, and distant stars, for discovering new phenomena whose existence we did not previously suspect, and for extending our investigation of nature. But in the same straightforward sense in which they share these beliefs, the realist holds many further beliefs about those entities, events, and phenomena that the instrumentalist declines to share: that the rocks are made up of atoms with a specific internal composition, that the spiders share a common ancestor with her in the distant past, that the path of the light from the distant stars can be bent by a gravitational field. Moreover, the scientific realist believes that there are entities in the world answering to the detailed descriptions given by her best scientific theories of entities like gluons, genes, and dark energy; by contrast, the instrumentalist believes that although these theoretical descriptions are intimately involved in the most powerful conceptual resources we have for thinking about particular natural domains, this does not give her a compelling reason to believe that there are entities in the world answering to those descriptions.

Furthermore, the realist and instrumentalist differ perhaps most radically and most fundamentally in their conceptions of and expectations concerning the future of scientific inquiry itself. The realist supposes that our further inquiry into the natural world will continue to bear out at least in broad strokes the various conceptions of domains of nature that are articulated by our current scientific theories. But the instrumentalist offers us a very different picture of the future of human scientific inquiry. She judges it quite likely that even the most genuinely impressive and instrumentally accomplished theories of contemporary science will ultimately be replaced by more powerful conceptual tools offering fundamentally different conceptions of nature that have presently not yet even been conceived.[13] She rejects the idea that even the most fundamental claims of theoretical science will persist indefinitely into the future as part of the best collection of conceptual tools we have for engaging the natural world. And among sensible people who are rightly impressed by the dramatic empirical successes of the best scientific theories we have, what greater difference could there be?

Notes

1. Moreover, there is reasonable controversy over classifying either Duhem or Poincaré as ultimately an instrumentalist in any of these senses: in his last work Poincaré reversed himself and wholeheartedly embraced the reality of atoms, while Duhem consistently held that scientific theories are able to establish "natural classifications" of the phenomena (for balanced discussion, see Psillos 1999 chap. 2, McMullin 1990, and Worrall 1982).

2. This historical progression is described in somewhat greater detail in Stanford (2005).

3. Fine himself takes a very different view of instrumentalism, insisting that the instrumentalist treats reliability (rather than truth) as the epistemic aim of *all* beliefs, and that she therefore need not specify what her instrumentalist commitment to a particular theory leads her to believe to be strictly and literally true: her commitment to *any* particular belief is simply a commitment to its reliability, full stop. Of course, if the "reliability" of our instrumentally reliable beliefs is not a matter of the true consequences they have, it seems natural to wonder just what such reliability consists in and how we are to characterize the various things that we think our instrumentally reliable beliefs allow us to do so reliably. But even setting this problem aside, embracing Fine's version of instrumentalism would seem a disproportionate response to the epistemic predicament we have considered in the preceding chapters. Perhaps it is *open* to us to be instrumentalists about all of our beliefs, just as it is open to us to be Cartesian skeptics, but this seems neither a warranted nor a measured response to the specific challenge

posed by the problem of unconceived alternatives: instead it offers yet another way for us to resemble the cat Twain describes in the passage that opens this chapter. Indeed, if the instrumentalist ultimately holds that the mere reliability of our best scientific theories gives them precisely the same epistemic status as every other successful belief we hold, it would seem that the concerns instrumentalists have traditionally raised regarding the distinctive epistemic status of our scientific theories are refuted rather than vindicated. In that event, we might even imagine that Fine's instrumentalist, far from opposing the realist's position on the epistemic status of scientific theories, is instead suggesting what the scientific realist might (coherently) have meant by insisting that they are (probably, approximately) true all along.

4. Of course this is perhaps ultimately a matter of preferring some perceptual judgments (e.g. those made with the aid of a ruler) to others on the grounds that they cohere better with the rest of our experience as a whole (e.g., they give the same results on repeated measurements, distinct agents arrive at the same values for them, etc.), but instrumentalism poses no obstacle to endorsing this preference.

5. Of course this means that specific elements of the hypothesis of the bodies of common sense have proved to be mistaken as well, but no version of it has ever been abandoned for a *fundamentally* distinct competitor, nor has history continuously revealed the sort of *fundamentally* different and previously unconceived alternatives that we find in the case of our scientific theories.

6. Likewise, there will be no such creature as "the Instrumentalist," who adopts an instrumentalist attitude toward any and all theories; there are instead only instrumentalists about particular theories. But I will continue to use the expression "the instrumentalist" as a shorthand description of a person who has adopted an instrumentalist attitude toward some particular theory or toward a substantial proportion of our fundamental scientific theories quite generally.

7. Furthermore, because we are not raising a special epistemic challenge that applies (by fiat) only to scientific theories as such, we need not pursue the dubious project of identifying just which theories are the "scientific" (i.e., suspect) ones. We have simply identified a perfectly general challenge for all theorizing about and theoretical knowledge of the world, though we recognize that the available evidence is sometimes (as in the case of the hypothesis of the bodies of common sense) sufficient to overcome the challenge. That is, the instrumentalist need not demarcate "scientific" theories from the rest, but can instead rest content with the conclusion that *particular theories* can overcome the quite general challenge developed here if they do not concern a subject matter with a demonstrated historical vulnerability to the problem of unconceived alternatives. But of course this will leave virtually all of our fundamental scientific theories vulnerable to the problem.

8. Of course, the realist will also say that what *makes* Newtonian mechanics useful is that for a certain range of phenomena its empirical implications approximate those of the true mechanics, but the instrumentalist certainly agrees with *that* claim (for more details, see Stanford 2000). Residual disagreement concerns whether the theory we have now is (probably, approximately) this true theory of mechanics or not.

9. Thus, to deny that our best scientific theories are probably and/or approximately true certainly need not and should not lead us to claim that they are therefore epistemically on a par with either rejected predecessors such as pangenesis or with such dramatically less empirically successful rivals as (God forbid) so-called creation science. From the instrumentalist's judgment that we are not epistemically entitled to belief in the central claims of a given scientific theory, it simply does not follow that every other theory is equally good, nor that we cannot judge particular alternatives to be completely unsupported or even definitively refuted by the available evidence.

10. Of course, if even our best theory of a given natural domain does *not* have such a record of impressive and fine-grained instrumental success to its credit the instrumentalist must be considerably more circumspect about believing what it implies in novel applications, but the sensible realist will surely share this more cautious attitude. Likewise, both the realist and the instrumentalist will be extremely cautious in applying even a successful theory to phenomena that are not systematically related to those with respect to which the theory's instrumental effectiveness has already been demonstrated.

11. This is emphatically *not* to insist that that Weismann's theory was the *only* justified option or that others (such as de Vries) might not have been equally justified in approaching practical contexts of prediction and intervention with a different set of conceptual tools. One consequence of accepting the form of epistemic instrumentalism presented here would be to undermine the insistence that only a single account of a given natural domain can be justified at any given time (cf. Kitcher 1993).

12. The rationale for this last demand may not be entirely obvious, as instrumentalism seems to offer little epistemic motivation for insisting that our theories not contradict one another, at least in those claims we do not believe. And indeed, the instrumentalist's motivation for consistency is pragmatic rather than epistemic: inconsistency between two theories is not simply further evidence against (at least one of) them, but rather threatens to prevent us from forming a single coherent set of practical expectations concerning any set of entities, events, or phenomena to which both apply. (Of course the instrumentalist would seem better positioned than the realist to make sense of and live with the fact that our best theories seem at present to be neither fully mutually consistent nor maximally unified.) Indeed, the demand that we try to unify our beliefs about the world into a coherent and consistent whole is in large part responsible for the recognition by both Kuhn and Laudan (see above) that our puzzles and problems may be theoretical or conceptual in character as well as empirical or narrowly practical. And as any number of theorists have emphasized (e.g.,Worrall 1982, Sklar 2000), seeking such consistency among our beliefs is a delicate matter of reciprocal influence and reflective equilibrium even between those of higher and lower levels of generality, neither subjugating our specific empirical accounts of nature to our more general metaphysical beliefs nor vice versa.

13. For this reason, we cannot defend realism as some thinkers have proposed to do (e.g., Ellis 1985) by embracing a Pragmatic, coherence, or "internal realist" conception of truth: one which simply identifies the truth about the world with whatever set of beliefs about it will or would emerge at the end of our scientific investigations or which identifies truth with warranted rational

assertibility in the limit of inquiry. This may well be the right view of truth to hold in any case, but it will simply not help to protect realism against the threat of unconceived alternatives, because the latter suggests that the accounts of nature embraced in the future course of our own theorizing about nature will be fundamentally and radically different from those of the present day.

References

Achinstein, Peter. (2002). "Is There a Valid Experimental Argument for Scientific Realism?" *Journal of Philosophy 99*: 470–495.

Barrett, Jeffrey A. and P. Kyle Stanford. (2005). "Prediction," in *The Philosophy of Science: An Encyclopedia*. Eds. Sahotra Sarkar and Jessica Pfeifer. New York: Routledge.

Black, J. (1803). *Lectures on the Elements of Chemistry, vol. 1*. Revised and prepared for publication in 2 volumes by John Robison, ed. Edinburgh.

Blackburn, Simon. (1984). *Spreading the Word*. Oxford: Clarendon Press.

Bowler, Peter J. (1989). *The Mendelian Revolution: The Emergence of Hereditarian Concepts in Science and Society*. Baltimore: Johns Hopkins University Press.

Boyd, Richard (1973). "Realism, Underdetermination, and a Causal Theory of Evidence," *Noûs 7*: 1–12.

———. (1981). "Scientific Realism and Naturalistic Epistemology." In *PSA 1980*, vol. 2. Eds. P. D. Asquith and T. Nickles. East Lancing, MI: Philosophy of Science Association.

———. (1984). "The Current Status of Scientific Realism." In *Scientific Realism*. Ed. Jarret Leplin. Berkeley: University of California Press, 41–82.

———. (1990). "Realism, Approximate Truth and Philosophical Method." Reprinted in *The Philosophy of Science*, 215–255. Ed. D. Papineau. New York: Oxford University Press, 1996.

Bulmer, Michael (2004). "Did Jenkin's Swamping Argument Invalidate Darwin's Theory of Natural Selection?," *British Journal of the History of Science 37*: 281–297.

Butterfield, Lisa H., Stephen P. Schoenberger, and Jeffrey B. Lyczak. (2004). "Cancer and the Immune System." In *Immunology, Infection, and Immunity*, ch. 24. Eds. Gerald B. Pier, Jeffrey B. Lyczak, and Lee M. Wetzler. Washington, D.C.: ASM Press.

Callender, L. A. (1988). "Gregor Mendel—An Opponent of Descent With Modification," *History of Science 26*: 41–75.

Campbell, Neil A. and Jane B. Reece. (2002). *Biology*, 6th Edition. San Francisco: Benjamin Cummings.

Christie, J. R. R. (1982). "Joseph Black and John Robison," in A. D. C. Simpson (ed.), *Joseph Black: 1728–1799*. Edinburgh: Royal Scottish Museum, 47–52.

Churchill, Frederick B. (1968). "August Weismann and a Break from Tradition," *Journal of the History of Biology 1*: 91–112.

———. (1970). "Hertwig, Weismann and the Meaning of Reduction Division Circa 1890," *Isis 61*: 428–457.

———. (1986). "Weismann, Hydromedusae and the Biogenetic Imperative, a Reconsideration," in Horder, Witkowski and Wylie (1986), pp. 7–33.

———. (1987). "From Heredity Theory to *Vererbung*: The Transmission Problem, 1850–1915," *Isis 78*: 337–64.

Coleman, William. (1965). "Cell, Nucleus, and Inheritance: An Historical Study," *Proceedings of the American Philosophical Society 109*: 124–158.

———. (1973). "Limits of the Recapitulation Theory: Carl Friedrich Kielmeyer's Critique of the Presumed Parallelism of Earth History, Ontogeny, and the Present Order of Organisms," *Isis 64*: 341–350.

Cowan, Ruth Schwartz. (1985). *Sir Francis Galton and the Study of Heredity in the Nineteenth Century*. New York: Garland Publishing.

Darwin, Charles. (1868). *The Variation of Animals and Plants Under Domestication* (1st American edition), 2 vols. New York: Orange Judd.

———. (1903). *More Letters of Charles Darwin*, 2 vols. Eds. Francis Darwin and A. C. Seward. London: John Murray.

———. (1905 [1874]). *The Variation of Animals and Plants Under Domestication*, 2d edition, 2 vols. London: John Murray. (1st ed. 1868.)

———. (1959 [1887]). *The Life and Letters of Charles Darwin*, 2 vols. Ed. Francis Darwin. New York: Basic Books.

De Vries, H. ([1889] 1910). *Intracellular Pangenesis*. Trans. C. Stuart Gager. Chicago: Open Court.

Donovan, A. (1993). *Antoine Lavoisier: Science, Administration, and Revolution*. Cambridge, MA: Blackwell Publishers.

Driesch, Hans. (1894). *Analytische Theorie der organischen Entwicklung*. Leipzig: Wilhelm Engelmann.

Duhem, Pierre. ([1914] 1954). *The Aim and Structure of Physical Theory*. Trans. from 2nd ed. by Philip P. Wiener. Originally published as *La Théorie Physique: Son Objet, et sa Structure* (Paris: Marcel Rivièra & Cie). Princeton, N.J.: Princeton University Press.

Dunn, L. C. (1965). *A Short History of Genetics*. New York: McGraw-Hill.

Earman, John. (1992). *Bayes or Bust? A Critical Examination of Bayesian Confirmation Theory*. Cambridge, MA: MIT Press.

———. (1993). "Underdetermination, Realism, and Reason," *Midwest Studies in Philosophy 18*: 19–38.

Ellis, Brian. (1985). "What Science Aims To Do," in Churchland and Hooker (eds.), *Images of Science*. Chicago: University of Chicago Press.

Endersby, Jim. (2003). "Darwin on Generation, Pangenesis, and Sexual Selection," in J. Hodge and G. Radick (eds.), *The Cambridge Companion to Darwin*. Cambridge: Cambridge University Press, 69–91.

Fine, Arthur. (1986). "Unnatural Attitudes: Realist and Instrumentalist Attachments to Science," *Mind 95*: 149–179.

———. (1991). "Piecemeal Realism." *Philosophical Studies* 61: 79–96.

Fitzgerald, George Francis. (1902). *The Scientific Writings of the Late George Francis Fitzgerald*. Ed. Joseph Larmor. Dublin: Hodges, Figgis, & Company.

Fix, John. (2004). *Astronomy: Journey to the Cosmic Frontier*, 3rd ed. New York: McGraw Hill.

Galton, Francis. (1865). "Hereditary Talent and Character," *McMillan's Magazine 12*: 157–166, 318–327.

———. (1869). *Hereditary Genius: An Inquiry into Its Laws and Consequences*. London: Macmillan.

———. (1870–1871). "Experiments in Pangenesis by Breeding from Rabbits of a Pure Variety, into Whose Circulation Blood taken from Other Varieties had Previously been Largely Transfused," *Proc. Roy. Soc. Lond. 19*: 394–410.

———. (1871–1872). "On Blood-Relationship," *Proceedings of the Royal Society of London 20*: 394–402.

———. (1875). "A Theory of Heredity," *Contemporary Review 27*: 80–95; revised in *J. Roy. Anthro. Inst. 5* [1876]: 329–348.

———. (1889). *Natural Inheritance*. London: Macmillan. Reprinted in Daniel N. Robinson, ed. *Significant Contributions to the History of Psychology 1750–1920, Series D, vol. IV: Darwinism*. Washington, D.C.: University Publications of America, 1977.

———. (1897). "The Contribution of Each Several Ancestor to the Total Heritage of the Offspring," *Proceedings of the Royal Society of London 61*: 401–413.

Gasking, Elizabeth B. (1967). *Investigations Into Generation: 1651–1828*. Baltimore: Johns Hopkins University Press.

Gayon, Jean. (1998). *Darwinism's Struggle for Survival: Heredity and the Hypothesis of Natural Selection*. Cambridge: Cambridge University Press.

Geison, Gerald. (1969). "Darwin and Heredity: The Evolution of His Hypothesis of Pangenesis," *Journal of the History of Medicine 24*: 375–411.

Glymour, Clark. (1977). "The Epistemology of Geometery," *Nous 11*: 227–251.

Gould, Stephen Jay. (1977). *Ontogeny and Phylogeny*. Cambridge, MA: Harvard University Press.

Guerlac, H. (1976). "Chemistry as a Branch of Physics: Laplace's Collaboration with Lavoisier." *Historical Studies in the Physical Sciences* 7: 193–276.

Haeckel, Ernst. (1866). *Generelle Morphologie der Organismen*, 2 vols. Berlin: Georg Reimer.

Hardin, C. and Alexander Rosenberg. (1982). "In Defense of Convergent Realism," *Philosophy of Science 49*: 604–615.

Hertwig, Oscar. (1896). *The Biological Problem of To-Day, Preformation or Epigenesis? The Basis of a Theory of Organic Development*. Trans P. Chalmers Mitchell. London: William Heinemann.

Hodge, M. J. S. (1985). "Darwin as a Lifelong Generation Theorist." in *The Darwinian Heritage*. Ed. D. Kohn. Princeton, N.J.: Princeton University Press, 207–243.

———. (1989). "Generation and the Origin of Species (1837–1937): A Historiographical Suggestion," *British Journal for the History of Science 22*: 267–281.

Hoefer, Carl and Alexander Rosenberg. (1994). "Empirical Equivalence, Underdetermination, and Systems of the World," *Philosophy of Science 61*: 592–608.

Horder, T. J., J. A. Witkowski, and C. C. Wylie, eds. (1986). *A History of Embryology*. Cambridge: Cambridge University Press.

Horwich, Paul. (1991). "On the Nature and Norms of Theoretical Commitment," *Philosophy of Science 58*: 1–14.

Kitcher, Philip. (1993). *The Advancement of Science*. New York: Oxford University Press.

———. (2001a). "Real Realism: The Galilean Strategy," *Philosophical Review 110*: 151–197.

———. (2001b). *Science, Truth and Democracy*. New York: Oxford University Press.

Kuhn, T. S. ([1962] 1996). *The Structure of Scientific Revolutions*, 3d. edition. Chicago: University of Chicago Press.

Kukla, Andre. (1993). "Laudan, Leplin, Empirical Equivalence and Underdetermination," *Analysis 53*: 1–7.

———. (1996). "Does Every Theory Have Empirically Equivalent Rivals?" *Erkenntnis 44*: 137–166.

Larson, J. L. (1979). "Vital Forces: Regulative Principles or Constitutive Agents?" *Isis 70*: 235–249.

Laudan, L. (1981). "A Confutation of Convergent Realism." Reprinted in *The Philosophy of Science*, 107–138. Ed. Papineau. New York: Oxford University Press, 1996.

———. (1984a). "Discussion: Realism without the Real," *Philosophy of Science 51*: 156–162.

———. (1984b). *Science and Values*. Berkeley: University of California Press.

———. (1984c). "Explaining the Success of Science: Beyond Epistemic Realism and Relativism." In *Science and Reality*. Eds. James Cushing, C. F. Delaney, and Gary Gutting. South Bend, IN: University of Notre Dame Press.

Laudan, Larry and Jarrett Leplin. (1991). "Empirical Equivalence and Underdetermination," *Journal of Philosophy 88*: 449–472.

Lavoisier, A. (1965 [1743–1794]). *Oeuvres de Lavoisier*, 6 vols., vols. 1–4. Ed. J. B. Dumas, vols. 5–6; ed. Edouard Grimaux, Imprimerie Impériale, Paris, 1862–93. New York: Johnson Reprint Corporation.

Lavoisier, A. and P. de la Place. (1982 [1783]). *Memoir on Heat*. Trans. and intro. Henry Guerlac. New York: Neale Watson Academic Publications.

Lenoir, T. (1981). "Teleology Without Regrets. The Transformation of Physiology in Germany: 1790–1847," *Studies in the History and Philosophy of Science 12*: 293–354.

———. (1982). *The Strategy of Life: Teleology and Mechanics in Nineteenth-Century German Biology*. Dordrecht: D. Reidel.

Leplin, Jarrett. (1997). *A Novel Defense of Scientific Realism*. New York: Oxford University Press.

Leplin, Jarrett and Larry Laudan. (1993). "Determination Undeterred: Reply to Kukla," *Analysis 53*: 8–16.

Magnus, P. D. and Craig Callender. (2004). "Realist Ennui and the Base Rate Fallacy," *Philosophy of Science 71*: 320–338.

Maienschein, Jane. (1986). "Preformation or New Formation—Or Neither or Both?" In Horder, Witkowski and Wylie (1986), pp. 73–108.

Maxwell, James Clerk. (1955 [1873]). *A Treatise on Electricity and Magnetism*, 2 vols. London: Oxford University Press.

Mazzolini, Renato G. and Shirley A. Roe. (1986). *Science Against the Unbelievers: The Correspondence of Bonnet and Needham, 1760–1780* (vol. 243 *of Studies on Voltaire and the Eighteenth Century*). Oxford: Alden Press.

McKie, D. (1935). *Antoine Lavoisier: The Father of Modern Chemistry*. London: Victor Gollancz.

McKie, D. and N. H. de V. Heathcote. (1935). *The Discovery of Specific and Latent Heats*. London: E. Arnold. Reprinted by Arno Press. New York: 1975.

McMullin, Ernan. (1990). "Duhem's Middle Way," *Synthese 83*: 421–430.

Morris, R. J. (1972.) "Lavoisier and the Caloric Theory." *British Journal for the History of Science* 6: 1–38.

Musgrave, Alan. (1988.) "The Ultimate Argument for Scientific Realism." In *Relativism and Realism in Science*. Ed. Robert Nola. Dordrecht: Kluwer Academic Publishers, 229–252.

Nagel, Ernst. (1961). *The Structure of Science*. New York: Harcourt, Brace, and World.

Norton, John D. (1993). "The Determination of Theory by Evidence: The Case for Quantum Discontinuity 1900–1915," *Synthese* 97: 1–31.

———. (1995). "Eliminative Induction as a Method of Discovery: How Einstein Discovered General Relativity." In *The Creation of Ideas in Physics: Studies for a Methodology of Theory Construction*. J. Leplin, ed. Dordrecht: Kluwer Academic Publishers, 29–69.

———. (2000). "How We Know About Electrons," in Robert Nola and Howard Sankey, eds., *After Popper, Kuhn and Feyerabend: Recent Issues in Theories of Scientific Method*. Dordrecht: Kluwer Academic Publishers, 67–97.

Olby, Robert C. (1963). "Charles Darwin's Manuscript of Pangenesis," *British Journal of the History of Science* 8: 85–93.

———. (1966). *Origins of Mendelism*, 1st Edition. New York: Schocken Books.

———. (1979). "Mendel no Mendelian?" *History of Science* 17: 53–72.

———. (1985). *Origins of Mendelism*, 2nd ed. (1st ed. 1966). Chicago: University of Chicago Press.

———. (1987). "William Bateson's Introduction of Mendelism to England: A Reassessment," *British Journal of the History of Science* 20: 399–420.

Pearson, Karl. (1914–1930). *The Life, Letters and Labours of Francis Galton, 3 vols*. Cambridge: Cambridge University Press.

Poincaré, Henri. ([1905] 1952). *Science and Hypothesis*. Reprint of first English translation; originally published as *La Science et L'Hypothèse* (Paris, 1902). New York: Dover.

Popper, Karl R. (1963). *Conjectures and Refutations*. London: Routledge and Kegan Paul.

Provine, William B. (1971). *The Origins of Theoretical Population Genetics*. Chicago: University of Chicago Press.

Psillos, Stathis. (1999). *Scientific Realism: How Science Tracks Truth*. New York: Routledge.

Putnam, Hilary. (1975). *Mathematics, Matter and Method (Philosophical Papers, vol. I)*, London: Cambridge University Press.

———. (1978). *Meaning and the Moral Sciences*. London: Routledge and Kegan Paul.

Quine, W. V. O. (1955). "Posits and Reality," Reprinted in *The Ways of Paradox and Other Essays*, 2nd ed. Cambridge, MA: Harvard University Press, 1976.

———. (1975). "On Empirically Equivalent Systems of the World," *Erkenntnis* 9: 313–328.

Robinson, Gloria. (1979). *A Prelude to Genetics: Theories of a Material Substance of Heredity, Darwin to Weismann*. Lawrence, KS: Coronado Press.

Roe, Shirley A. (1981). *Matter, Life, and Generation: Eighteenth-Century Embryology and the Haller-Wolff Debate*. New York: Cambridge University Press.

Salmon, Wesley. (1990). "Rationality and Objectivity in Science or Tom Kuhn Meets Tom Bayes." Reprinted in *The Philosophy of Science*. Ed. D. Papineau. New York: Oxford University Press, 1996.

Shimony, Abner. (1970). "Scientific Inference." In *The Nature and Function of Scientific Theories*, vol. 4 of the University of Pittsburgh series in the Philosophy of Science. Ed. Robert G. Colodny. Pittsburgh: University of Pittsburgh Press.

Sklar, L. (1975). "Methodological Conservatism," *Philosophical Review 84*: 384–400.

———. (1981). "Do Unborn Hypotheses Have Rights?" *Pacific Philosophical Quarterly* 62: 17–29.

———. (1982). "Saving the Noumena," *Philosophical Topics 13*: 49–72.

———. (2000). *Theory and Truth*. New York: Oxford University Press.

Smart, J. J. C. (1968). *Between Science and Philosophy*. New York: Random House.

Spencer, Herbert. (1864). *Principles of Biology*, 2 vols. London: Williams and Norgate.

Stanford, P. Kyle. (2005). "Instrumentalism," in *The Philosophy of Science: An Encyclopedia*. Eds. Sahotra Sarkar and Jessica Pfeifer. New York: Routledge.

———. (2000). "An Antirealist Explanation of the Success of Science," *Philosophy of Science 67*: 266–284.

Stanford, P. Kyle and Philip Kitcher. (2000). "Refining the Causal Theory of Reference for Natural Kind Terms." *Philosophical Studies 97*: 99–129.

Stein, Howard. (1989). "Yes, but... Some Skeptical Remarks on Realism and Anti-Realism," *Dialectica 43*: 47–65.

Van Fraassen, Bas. (1980). *The Scientific Image*. Oxford: Oxford University Press.

———. (1985). "Empiricism in the Philosophy of Science." In *Images of Science*. Eds. P. M. Churchland and B. A. Hooker. Chicago: University of Chicago Press.

———. (1989). *Laws and Symmetry*. Oxford: Oxford University Press.

Von Baer, Karl Ernst. (1828 [1951]). *Uber Entwickelungs-geschichte der Thiere*. Trans. T. H. Huxley as *On the Development of Animals, with Observations and Reflections*. Excerpted in *A Source Book in Animal Biology*. Ed. Thomas S. Hall. New York: McGraw-Hill, 392–399.

Weismann, August. (1883). "On Heredity." Reprinted in Weismann (1891–2), vol. I, pp. 67–106.

———. (1885). "The Continuity of the Germ-Plasm as the Foundation of a Theory of Heredity." Reprinted in Weismann (1891–1892), vol. I, pp. 163–254.

———. (1887). "On the Number of Polar Bodies and their Significance in Heredity." Reprinted in Weismann (1891–1892), vol. I, pp. 343–396.

———. (1890). "Remarks Upon Certain Problems of the Day." Reprinted in Weismann (1891–2), vol. II, pp. 71–97.

———. (1891–2). *Essays Upon Heredity and Kindred Biological Problems*, vol. 1. Trans. and eds. Edward B. Poulton, Selmar Schönland, and Arthur E. Shipley; Eds. and trans. Edward B. Poulton, Arthur E. Shipley, vol. II. Oxford: Clarendon Press.

———. ([1892] 1893). *The Germ-Plasm, A Theory of Heredity*. Trans. W. Newton Parker and Harriet Rönnfeldt. New York: Charles Scribner's Sons.

———. (1896). *On Germinal Selection as a Source of Definite Variation*, Trans. T. J. McCormack. Chicago: Open Court.

Winther, Rasmus G. (2000). "Darwin on Variation and Heredity," *Journal of the History of Biology 33*: 425–455.

Worrall, J. (1989). "Structural Realism: The Best of Both Worlds?," *Dialectica 43*: 99–124.

———. (1994). "How to Remain (Reasonably) Optimistic: Scientific Realism and the 'Luminiferous Ether.'" In *PSA 1994*, vol. 1. Eds. D. Hull, M. Forbes and R. M. Burian. East Lansing, MI: Philosophy of Science Association.

Zumdahl, Steven S. and Susan Arena Zumdahl. (2003). *Chemistry*, 6th ed. Boston: Houghton Mifflin.

Index